SHORE WILDFLOWERS OF
CALIFORNIA, OREGON, AND WASHINGTON

Shore Wildflowers of California, Oregon and Washington

by Philip A. Munz

UNIVERSITY OF CALIFORNIA PRESS

Berkeley, Los Angeles, London

UNIVERSITY OF CALIFORNIA PRESS
Berkeley and Los Angeles, California

UNIVERSITY OF CALIFORNIA PRESS, LTD.
London, England

ISBN: 0-520-00903-7
Library of Congress Catalog Card Number 64-22585
Printed in the United States of America

3 4 5 6 7 8 9 0

CONTENTS

INTRODUCTION

This is the fourth of the so-called "wildflower books" dealing with California plants, but in this case there is a wider geographical range, namely all three Pacific states. The original and still the main purpose of these little books is to enable the layman interested in knowing something about the plants he finds growing wild, to get some notion as to what these plants are called, how they may be recognized, where they are to be sought, and what their general relationship may be. However, interestingly enough, it has developed that many of the first three books are being purchased by students and others who use the more technical volume, A *California Flora*, by Munz and Keck, in order to find illustrations and to supplement the usability of that volume. This fourth wildflower book brings to more than one thousand the number of species illustrated.

I have always thought, too, and now with such student use to be kept in mind I feel it increasingly valid, that many a layman may wish to know not only the pretty and conspicuous plants. Often others excite his curiosity. This is especially true in our coastal salt marshes where many forms grow which are striking by virtue of peculiar structure: fleshiness, jointed stems, odd-looking inflorescences, and the like. On coastal dunes there may pop up out of the sand root-parasites lacking chlorophyll, not showy, but displaying intriguing small purple flowers with white rims. The layman may want to know the difference between a grass, a sedge, and a rush, all more or less alike in their small flowers and often conspicuous as dune-binders and marsh inhabitants. Since the number of kinds of trees that come down to the shore and its adjacent bluffs is limited, it seems well too to include as many of them as space permits. So, although some of my reviewers in the past have criticized the selection of species used for my wildflower books and the space given to inconspicuous species, I must have in mind all the readers to whom I hope the books may be of use.

While I have selected the areas covered by the four volumes (*California Spring Wildflowers, California Desert Wildflowers, California Mountain Wildflowers* and now *Shore Wildflowers of California, Oregon, and Washington*) in such fashion as to have quite different species represented in the four, a small amount of duplication is inevitable. It is, however, so little, that it can almost be disregarded.

1

Shore Climate

As explained in the first paragraph, this book deviates from the pattern previously chosen in that the earlier volumes were more definitely limited to the state of California. But as one studies beach and shore species, he is immediately aware of the fact that many grow far to the north, even to Alaska. The cooler, more regular, climate of the coast as compared with that of the interior and the longer lasting effect of even small amounts of rain along the coast as compared with inland valleys, make it obvious that the same plant may range, if not from southern California, at least from central California far into Oregon or Washington, or even British Columbia or Alaska. Coastal summers are relatively cool. Coastal winters are relatively warm. Hence, the present volume attempts to help in identification of not only California shore wildflowers, but those of Oregon and Washington. In fact, some species are included from these two latter states that do not range south into California or only into its extreme northern part.

What is the Shore?

The first question that arose in my mind when I was asked to write a book on shore wildflowers was: what should the contents be? Naturally, the actual sandy beach and dunes, which together might be called the coastal strand, would come first. Next would be included the coastal salt marshes with their rather distinctive flora. Then I would include the bluffs along the coast, especially as far inland as salt spray seems to influence. Such influence is evident in southern California, particularly in the appearance of desert plants like Bladderpod (*Isomeris arborea*), of more or less saline conditions. But as one goes north, this influence decreases, although in much of central California there is a well-marked zone of coastal scrub, with quite different plants than in the redwood or other forests behind it. I take it that this coastal scrub is still due, at least in part, to the influence of salt spray. But still farther north with still greater rainfall, as in extreme northern California and in Oregon and Washington, the forests come right down to the bluffs and to the actual sand of the back beaches. I believe that with the high rainfall in this area, any possible effect of salt is almost immediately leached out and that the actual shore is invaded by plants like Pearly Everlasting (*Anaphalis margaritacea*) a Giant Horsetail (*Equisetum Telmateia*) normally of wooded places, whereas a little to the south species like False Lily-of-the-Valley (*Maianthemum dilatatum*) come out on to the actual beach only along freshwater streams. In some ways, then, the northern coast has more species that normally grow in the adjacent forests than does the southern. Then, too, with the greater rainfall in the north, sandy areas are more easily taken over by Spartina and other perennials and there is not the development of as rich a strand flora, for the most part, as there is in Monterey and San Luis Obispo counties of California, although a possible exception might be cited at Gold Beach, Oregon.

CHARACTERISTICS OF SHORE PLANTS

Apparently the most important single factor in the environment of shore plants which sets them apart from those farther inland is the presence of salt or salts in the soil. Dissolved salts mean physiological dryness for the plant, which then has to contain within itself a higher percentage of dissolved substances to pull in water by osmosis than it would if in pure water. This is true whether growing in arid regions like the desert, where the dissolved salts in the soil may be appreciable and where they may even coat the surface with a layer of so-called alkali, or whether found along the coast.

Usually plants of these two types of environment have a reduced evaporating surface as compared with those in what is called a mesophytic environment with abundance of good water in the soil, as in the garden or in a region of high rainfall. Reduction of evaporating surface may mean the development of thickened fleshy leaves or the replacement of functional leaves by fleshy green stems in which the process may occur. In either case, there is a resultant succulence and this is noticeable to the person acquainted with the plants of an inland environment who finds himself at the beach. He may run onto species closely related to those with which he is familiar, but quite different in their succulence and compactness, or he may find more succulent forms of the same species. Examples are the Sand-Verbenas and Fiddleneck of the coast and of the interior.

Plants specially adapted to life in soils with high concentration of salts are often called *halophytes*. Good examples are some of the species of Saltbush (*Atriplex*), Salt Grass (*Distichlis*), Pickleweed (*Salicornia*), Seepweed (*Suaeda*), and the Sea-Fig and Hottentot-Fig (*Mesembryanthemum*). Some of these are on sandy strands, others in the coastal salt marshes. For the most part these halophytes are not beautiful, but they can be quite striking in appearance.

WHICH WILDFLOWERS ARE TO BE SOUGHT IN THIS BOOK?

I have used the term wildflower very loosely, as mentioned earlier in this introduction, making it almost synonymous with the word "plant," or perhaps better, "higher plant." For example, there are often cast up on the beaches with broken off pieces of kelp or seaweed, two flowering plants that grow entirely submerged, especially in shallow bays, namely Eel-Grass (*Zostera*) and Surf-Grass (*Phyllospadix*). These are included on page 79. Then I have given a small section to some of the coastal ferns and horsetails, yet they do not produce flowers. And we do not think of shrubs and trees as wildflowers, but a good many of them are discussed. In other words, I have attempted to treat in this little book the more interesting higher plants found near or on the shore, as well as the showy forms such as Azalea (*Rhododendron*), California Poppy (*Eschscholzia*), and Foxglove (*Digitalis*).

Of course the scope of a small book like this is limited and it cannot possibly contain all the plants to be found in the area under discussion. Often what is here available will indicate that the species you identify is a Gilia, a Phacelia, or a Castilleja, but perhaps not the one actually illustrated. For further and more exact information you are referred to the larger, more complete, volume by Munz and Keck, *A California Flora*, University of California Press, 1959, which is available in California stores and libraries. For farther north see Peck, *A Manual of the Higher Plants of Oregon*, Oregon State University Press, 1961, and Piper, *Flora of Washington*, a Government publication, 1906, and probably available only on the secondhand market or in public libraries.

How to Identify a Wildflower

To refresh the reader's memory a drawing is presented in figure A showing the principal parts of a typical flower. In the text of the book it seems impossible to discuss plants and their flowers without using

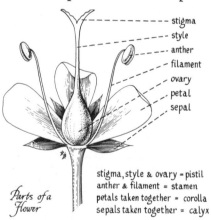

stigma, style & ovary = pistil
anther & filament = stamen
petals taken together = corolla
sepals taken together = calyx

Parts of a Flower

Figure A. A Representative Flower

the names of the parts. In the typical flower we begin at the outside with the *sepals*, which are usually green, although they may be of other colors. The *sepals* together constitute the *calyx*. Next comes the *corolla* made up of separate *petals*, or the petals may be grown together to form a tubular or bell-shaped or wheel-shaped corolla. Usually the corolla is the conspicuous part of the flower, but it may be reduced or be lacking altogether (as in grasses and sedges) and its function of attraction of insects and other pollinators may be assumed by the calyx. The calyx and corolla together are sometimes called the perianth, particularly where they are more or less alike. Next as we proceed inward in the flower, we usually find the *stamens*, each typically consisting of an elongate *filament* and a terminal *anther*, which latter produces the pollen. At the center of the flower are one or more *pistils*, each with a basal *ovary* containing the ovules or immature seeds, a more or less elongate *style*, and a terminal *stigma* with a rough sticky surface for catching pollen. In some species stamens and pistil are borne in separate flowers or even on separate plants. In the long evolutionary process by which plants have developed into the many diverse types of the present day and by which they have been adapted to various pollinating agents, their flowers have undergone very great modifications and so now we find more variation in the flower than in other plant parts. Hence, classification is largely dependent on the flower.

To help the reader identify a flower, color plates are given for 96 species, and drawings, when used, are grouped by color. In attempting to arrange plants by flower color, however, it is difficult to place a given species to the satisfaction of everyone. The range of color may vary so completely from deep red into purple, or from white to whitish to pinkish, or from blue into lavender, that it is impossible to satisfy the writer himself, let alone his readers. I have done the best I could and have tried for the general impression given as to color, especially when the flowers are minute and the general color effect may be caused by parts other than the petals. My hope is that by comparing a given wild-flower with the drawing it resembles under the color in which the reader feels it ought to be placed and then by checking with the facts given in the text, he may in most cases succeed in discovering what his plant may be.

<h3 align="center">ACKNOWLEDGMENTS</h3>

Most of the drawings for this book were made by Jeanne R. Janish (Mrs. Carl F. Janish), whose illustrations in books on western botany are so well known and so successful in recreating the appearance of a plant in three dimensions from a pressed specimen, that any botanical author feels proud to be able to say that Mrs. Janish is his illustrator. Her drawings in this book are always indicated by the letter "J." A few illustrations were used from an earlier work of mine long out of print and covering the plants of southern California. These were made by Tom Craig, Rodney Cross, and Milford Zornes, who were at that time undergraduate students at Pomona College where I was teaching. Each of their drawings is signed. Two drawings are the work of Dick Beas-ley: Fig. A. showing the parts of a representative flower, and figure

143, *Croton californicus*. In addition to those mentioned above some are scattered through the text without any identifying mark. They were made by Dr. Stephen Tillett: namely, figure 12, *Isomeris arborea;* 14, *Erysimum Menziesii;* 21, *Oenothera ovata;* 46, *Atriplex semibaccata;* 76, *Limonium californicum;* 114, *Salicornia subterminalis;* 135, *Rubus vitifolius;* 142, *Oxalis oregana;* 153, *Trientalis latifolia;* 156, *Linanthus grandiflorus;* and 159, *Cryptantha intermedia.*

Most of the Kodachromes are my own. Those obtained from others are listed below. Photographs for *Pholisma arenarium* and *Malacothrix incana* were kindly loaned by Dr. Sherwin Carlquist. Miss Beatrice F. Howitt generously donated several: *Eriogonum latifolium, Cakile maritima, Erysimum suffrutescens, Potentilla Egedei, Rubus vitifolius, Conium maculatum, Armeria maritima, Vaccinium ovatum,* and *Convolvulus Soldanella.* The remainder are from the collection of the Rancho Santa Ana Botanic Garden and were made as follows: *Mesembryanthemum crystallinum* by M. Carrothers; *Fragaria chiloensis* by Dr. R. J. Shaw; *Castilleja foliolosa* by Brooking Tatum; *Zauschneria cana* by Dr. R. F. Thorne; *Heliotropium curassavicum* by Dr. S. Tillett; *Ranunculus californicus, Ceanothus griseus* and *C. megacacarpus, Solidago occidentalis* by Freda Wertman; and *Sambucus callicarpa* by Dr. Carl B. Wolf. A much larger number were the work of Mr. P. C. Everett: *Iris Douglasiana, Eriogonum latifolium grande, E. giganteum, Dudleya pulverulenta, D. virens, Heuchera micrantha, Ribes speciosum, Holodiscus discolor, Limnanthes Douglasii, Arctostaphylos columbiana, Encelia californica, Coreopsis maritima, Lasthenia glabrata, Grindelia stricta.* I am most grateful to the various persons who have aided me by such kodachromes.

It is a pleasure, too, to thank James Henrickson for his careful lettering on the map of coastal counties of Washington, Oregon, and California. The original map was from a tracing that I made. This was inked by Mrs. Janish and the names were lettered by Mr. Henrickson.

Once again, I am gratefully indebted to Susan J. Haverstick, who has so understandingly edited the previous books: *A California Flora, California Spring Wildflowers, California Desert Wildflowers,* and *California Mountain Wildflowers.* Her suggestions on the text itself and her ideas have been most helpful. I thank her once more for participating in this final wildflower book on California and the other states on the Pacific Coast.

Philip A. Munz
Rancho Santa Ana Botanic Garden
Claremont, California
February 10, 1964

FERNS AND FERN ALLIES

Section One

Although not "wildflowers," the ferns
and their allies are prominent plants and
of interest to many. Furthermore, several
kinds grow on rocky sea bluffs and in
shaded and protected places right down
to the edge of the beach, especially
northward. So, for completeness, several
are here included, such as POLYPODY
(*Polypodium Scouleri*), figure 1, with a
creeping woody rootstock covered with a
loose chaff. The leaves are thick, almost
leathery, once divided into blunt segments, and four to sixteen inches high.
Their under surface bears naked round
fruit dots crowded against the midribs.
This fern is found on mossy logs, cliffs,
and slopes from Santa Cruz Island, California, to British Columbia.

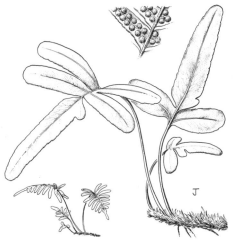

FIGURE 1.   POLYPODY

SILVERBACK FERN (*Pityrogramma tri-
angularis* var. *viscosa*), figure 2, is a
rather small fern from a stout ascending
or short-creeping rootstock covered with
brownish or blackish scales. The fronds
are triangular in shape, on red-brown
stipes or petioles, and sticky above and
white-powdery beneath. In the dry season they become quite curled up. The
spore-producing fruit dots are not conspicuous, being borne on the under surface and along the veins. Silverback
occurs on coastal slopes of San Diego
and Orange counties, California, and on
several of the California islands.

FIGURE 2.   SILVERBACK FERN

CALIFORNIA LACE FERN (*Aspidotis
californica*), figure 3, was formerly included in Cheilanthes. It is easily recog-
nizable by its finely divided triangular
leaf blades which are hairless and thickish in texture. They too curl up tightly in
dry weather. These fronds come from a
scaly short-creeping rootstock, have wiry
dark brown stipes, and reach a height of
three to twelve inches. The fruit dots or
fertile spots are on the under surface and

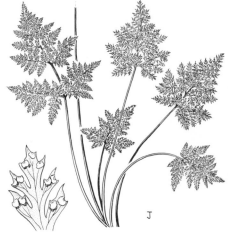

FIGURE 3.   CALIFORNIA LACE FERN

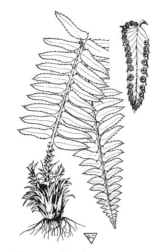

FIGURE 4.   SWORD FERN

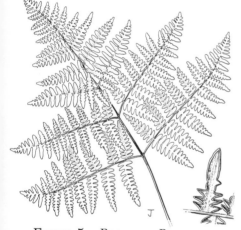

FIGURE 5.   BRAKE or BRACKEN

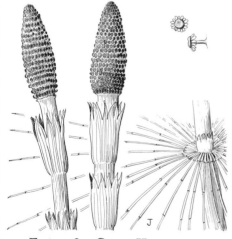

FIGURE 6.   GIANT HORSETAIL

protected by revolute patches on the margin. Lace Fern ranges about rocky places from northern Lower California to Humboldt County, California.

SWORD FERN (*Polystichum munitum*), figure 4, is a species of damp woods in cool places along the coast from Santa Cruz Island and Monterey County, California, north to Alaska. It is a coarse evergreen from woody, suberect, very scaly rootstocks. The fronds are many, in heavy crowns or clumps two to four feet high, with stout stipes (petioles) having conspicuous chestnut-brown scales. The leaves are once divided in pinnate fashion and on the under surface bear many rounded, submarginal fruit dots, each covered by a flap of tissue.

BRAKE or BRACKEN (*Pteridium aquilinum* var. *lanuginosum*), figure 5, has long-creeping, branched, underground rootstocks clothed with hairs and sending up stout fronds one to five feet tall. The stipes are straw-colored, the blades three-times divided and one to four feet long. On their under surface they may have an inrolled margin under which are borne the fruit dots. This fern is widely distributed in California and north to Alaska.

Distantly related to the ferns and reproducing like them by spores instead of by flowers, is the Horsetail or Scouring-Rush. GIANT HORSETAIL (*Equisetum Telmateia* var. *Braunii*), figure 6, has in the early spring short-lived unbranched stems which are whitish or brownish and jointed. At the tip of each joint is a membranous sheath with twenty to thirty teeth. The summit of the stems bears a kind of cone one to three inches long and producing the reproductive cells or spores. Indicated in the drawing is also a piece of the fertile stem that comes later in the season and is discussed on page 27.

FLOWERS YELLOW TO ORANGE

Section Two

Yellow Skunk-Cabbage (*Lysichiton americanum*), figure 7, like others of the Arum Family is characterized by having a greenish central fleshy spike or spadix made up of many minute flowers and surrounded by a modified, in this case yellowish, leaf or spathe. A spring bloomer, this plant has ill-smelling flowers and large rather fleshy leaves, and inhabits swampy or wet places along the coast from the Santa Cruz Mountains, California, north to Alaska. Familiar members of the family are Philodendron, Calla-Lily, and Jack-in-the-Pulpit.

Hottentot-Fig (*Mesembryanthemum edule*), figure 8, is a member of the Carpet-Weed Family and itself is an introduction from South Africa. It has been much planted along highways and banks to control erosion and has become naturalized on dunes and sandy places along the coast, where it grows with the so-called Sea-Fig (page 30). The latter has rose-magenta flowers instead of the yellowish ones of the former species. Both have very fleshy three-sided leaves that are somewhat curved in the Hottentot-Fig and straight in the Sea-Fig.

The Buttercup Family is important especially in temperate zones, the true Buttercup (see also pages 31 and 97) having shining yellow petals, many stamens, and several central small pistils, each of which matures into a minute one-seeded dry fruit. The form shown in figure 9 is *Ranunculus californicus* var. *cuneatus,* which is prostrate and grows on coastal bluffs and hills of Santa Cruz and San Miguel islands and along the coast from Monterey County, California, to Oregon. It is a small plant, the stems attaining a length of about ten inches. Other usually more inland and even

Figure 7.   Yellow Skunk-Cabbage

Figure 8.   Hottentot-Fig

Figure 9.   Buttercup

FIGURE 10. BUTTERCUP

FIGURE 11. BARBERRY or MAHONIA

FIGURE 12. BLADDERPOD

montane forms of *R. californicus* may be erect and two feet high.

Another coastal BUTTERCUP is *Ranunculus acris*, figure 10, native of the Old World and long since introduced into America and naturalized in moist places across the continent. It grows near the coast from Humboldt County, California, to Washington. It is a more or less hairy perennial with several erect stems and is one and one-half to three feet tall and branched above. The five yellow petals are one-third to one-half inch long; the more or less five sided leaves are two to three inches wide and deeply cut. The roots are stout and persist from year to year.

BARBERRY or MAHONIA (*Berberis pinnata*), figure 11, of the Barberry Family, is woody with prickly somewhat hollylike leaves. The flowers are built on the plan of three, the six petals being in two series. The stamens are six, with flattened filaments. The fruit is a bluish berry about one-fourth inch long and with acid sap. The wood and inner bark of the barberries are yellow and have been used in times past for dye. This species is one of several on the Pacific Coast, growing in rocky exposed places on Santa Cruz and Santa Rosa islands and the mainland of California and Oregon.

A shrub with an interesting distribution, namely deserts and coastal sea bluffs where conditions are somewhat saline, is BLADDERPOD (*Isomeris arborea*), figure 12, of the Caper Family. Everyone knows the Caper of cookery, an Old-World plant. Our American members of the family mostly have ill-smelling foliage, four-petaled flowers, and stalked, often inflated, pods. Isomeris is a shrub to several feet tall, with three-parted leaves and yellow flowers an inch or more across. Coastally it extends as far north as San Luis Obispo County, California.

The Mustard Family, like the Caper Family, has four-petaled flowers and peppery sap, as witnessed by the Radish, Cabbage, and other cultivated plants. One of its common wild representatives is WALLFLOWER, the one shown in figure 13 being *Erysimum franciscanum*. The flowers in this species are yellow to cream, over half an inch wide; the plant is a short-lived perennial to a foot or more high. The seed-pods are an inch or more long, erect, often tinged purple. This coastal species is distributed from San Mateo County, California, to southwestern Oregon and is similar to other local species on sand dunes and bluffs fairly extensively distributed all along the Pacific Coast (see page 32).

Another species of WALLFLOWER is *Erysimum Menziesii*, quite different in its habit, as can be seen in figure 14. With a long taproot and widely spreading seed-pods, as well as differently shaped broader leaves, it serves as an example of another type. The bright yellow flowers are fragrant as in other species. It occurs locally on dunes at Point Pinos in Monterey County, California, and also in Mendocino and Humboldt counties. In southern California other species are narrow leaved and tend to be more or less woody at the base of the plant.

In the Mustard Family too is FIELD MUSTARD (*Brassica campestris*), figure 15, an erect annual one to three feet tall. Characterized by its clasping upper leaves and bright yellow flowers, it is an attractive plant, albeit an introduced weed from Europe. Being an annual, it is not excessively pernicious. It is widely distributed inland, as in orchards and

FIGURE 13.   WALLFLOWER

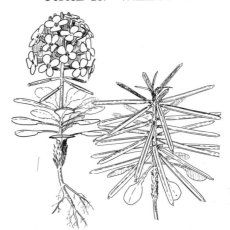

FIGURE 14.   WALLFLOWER

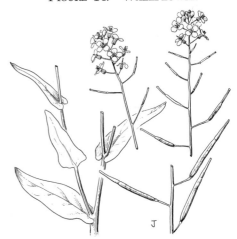

FIGURE 15.   FIELD MUSTARD

FIGURE 16. LOTUS

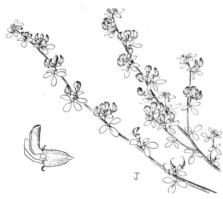

FIGURE 17. LOTUS

FIGURE 18. LOCOWEED

fields, but grows in sandy places and on bluffs along most of our coast, flowering in early spring in the south and into July farther north.

With flowers characteristic of the Pea Family the LOTUS (*L. salsuginosus*), figure 16, is a prostrate annual with smooth or slightly hairy stems to about one foot long and slightly fleshy leaves. The flowers are mostly yellow, becoming reddish with age, about one-third inch long, and with somewhat longer straight seed-pods. The species is found on the Channel Islands of California and in sandy places and on sea bluffs from Santa Clara County south, penetrating also into the interior. See also page 39.

Another coastal member of the genus Lotus is *L. junceus*, figure 17. This LOTUS is a slender-stemmed perennial, much branched, somewhat woody at the base, with fine appressed hairs and stems to twenty inches long. The corolla is about one-fourth inch long, yellow, tinged red, and the seed-pod is somewhat curved and beaked. It occurs on dry coastal hills from Mendocino County, California, south to San Luis Obispo County. See also page 39.

One of the large groups in western North America in the Pea Family is comprised of the Locoweeds or Rattleweeds, with well over one hundred species in the Pacific states. In some areas they cause considerable poisoning of livestock. A species of LOCOWEED with yellowish flowers is *Astragalus pycnostachyus*, figure 18, a stout perennial more or less woolly with short twisted hairs. The erect stems grow to almost three feet and bear large pinnate leaves with crowded leaflets. The flowers and seed-pods are about one-half inch long. This Astragalus occurs in salt marshes or moist depressions behind dunes along much of the California coast. See also page 97.

The genus Viola of the Violet Family (page 73) is a garden and a wild favorite, the most coastal species being the EVERGREEN VIOLET (*Viola sempervirens*), figure 19. It is an almost hairless perennial from short scaly rootstocks and it produces stolonlike stems with scattered rounded leaves and lemon-yellow flowers to one-half inch long. The three lower petals are faintly purple-veined. Growing mostly in shaded woods, it may occur right down to the edge of the beach and is distributed from central California to British Columbia.

The Cactus Family is not primarily one of the immediate coast, but several species do grow on coastal bluffs (see page 41). One of these is a PRICKLY-PEAR (*Opuntia littoralis*), figure 20. It is a large plant to four or five feet high and with more or less elongate joints a foot or so in length. The spines are whitish with red-brown base, the flowers are yellow. It is distributed from Santa Barbara County, California, to Lower California. With it grows another species with roundish joints and yellow spines. Other closely related plants occur inland.

Another yellow-flowered plant is SUN CUP (*Oenothera ovata*), figure 21, of the Evening-Primrose Family (see pages 42, 43, 67). It is four-petaled as in the Mustard and Caper Families, but the ovary or seed-bearing part is below instead of above the petals. In this species the flower is at the summit of a long tube and the ovary is hidden in the tuft of leaves at the base of the plant. Unlike many of the evening-primroses, the Sun Cup is a day bloomer with bright yellow flowers highly reminiscent of the buttercup. It is found in open places along the

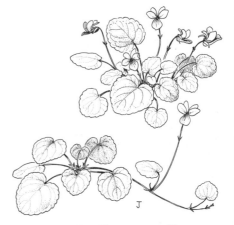

FIGURE 19.   EVERGREEN VIOLET

FIGURE 20.   PRICKLY-PEAR

FIGURE 21.   SUN CUP

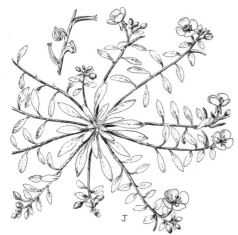

FIGURE 22. BEACH-PRIMROSE

FIGURE 23. YELLOW MATS

FIGURE 24. MOCK-AZALEA

coast from southern Oregon to San Luis Obispo County, California.

The BEACH-PRIMROSE (*Oenothera cheiranthifolia*), figure 22, is another Sun Cup in that it is a day bloomer. A perennial with more or less prostrate stems radiating from a central rosette of leaves, it is usually grayish-hairy throughout and forms large mats in full maturity. The yellow petals may turn red in age and may be one-fourth to two-thirds inch long. Growing on the coastal strand, it occurs from Coos County, Oregon, south. From Point Conception to northern Lower California it becomes more woody and has larger flowers than north of Santa Barbara County. See page 43.

YELLOW MATS (*Sanicula arctopoides*), figure 23, is a member of the Carrot Family, a group of aromatic herbs with small flowers arranged in umbels or clusters radiating from a given level. Parsley, Parsnip, Carrot, Dill, Coriander, and Anise are familiar examples. Yellow Mats is a more or less prostrate perennial with three-parted leaves and small yellow flowers producing bristly fruits. It is found in sandy flats and on open hillsides mostly near the coast from northern Oregon to central California. See pages 43, 44, 73, 99, 100.

MOCK-AZALEA (*Menziesia ferruginea*), figure 24, belongs to the Heath Family, together with Rhododendron, Manzanita, Blueberry, and others (pages 44–46, 101). It is a deciduous shrub to several feet high and has glandular-pubescent twigs. The leaves are one to two inches long and the flowers are yellow tinged with red. The fruit is a dry capsule one-fourth inch long. Growing along the coast of Humboldt and Del Norte counties, California, this species ranges to Alaska and Montana. It flowers in June and July.

A roadside weed near the coast from north central California to Oregon is PARENTUCELLIA (*P. viscosa*), figure 25, of the important Figwort Family known for plants such as Snapdragon, Paint-Brush, Monkey-Flower, and Penstemon (pages 49–51, 67, 68, 76). This glandular annual has opposite toothed leaves and terminal leafy spikes of yellow flowers with two-lipped corollas over half an inch long. It is abundant in disturbed places and is an introduction from the Mediterranean region. It blooms from April to June.

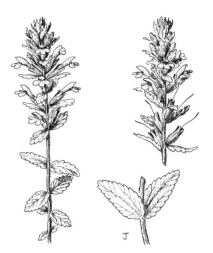

FIGURE 25.   PARENTUCELLIA

Another member of the Figwort Family is a native plant, JOHNNY NIP (*Orthocarpus castillejoides*), figure 26, which grows in low saline places and on sea bluffs from British Columbia to Monterey County, California. The corolla ranges from one-half to one inch long and is yellow with purple markings. About Humboldt Bay is a form (var. *humboldtiensis*) with a purplish corolla having the lower lip tipped with yellow. The illustration shows separately some of the bracts that subtend the flowers of this species.

FIGURE 26.   JOHNNY NIP

A coastal California BEDSTRAW is *Galium californicum*, figure 27, ranging from Humboldt to Los Angeles counties, California. It has slender creeping rootstocks and tufted slender, retrorse-hairy stems to about one foot long, which bear whorls of four leaves one-fourth to one-half inch long. The staminate flowers occur largely in groups of two or three; the pistillate flowers are solitary. The corolla is yellowish and very small. The fruit is fleshy, smooth or hairy, and becomes white when ripe, then blackens as it dries. It is of the Madder Family to which belongs the Gardenia!

The Sunflower Family (pages 52–58, 68, 76, 104) is noteworthy because of its

FIGURE 27.   BEDSTRAW

FIGURE 28.   HAPLOPAPPUS

FIGURE 29.   HAPLOPAPPUS

FIGURE 30.   GOLDEN-YARROW

great size. In it the many small florets or flowers are produced in a head subtended by a series of bracts forming an involucre and such a head tends to resemble a solitary flower of other families. The florets may all be alike and strap-shaped as in the common Dandelion, or all tubular as in the Pincushion Flower, or of both kinds as in HAPLOPAPPUS (*H. racemosus*), figure 28. This perennial, with a short taproot and a tuft of basal leaves, bears several stems to three feet long. The heads are several to many, the so-called ray-flowers (elongate and petal-like) to about one-half inch long. The species occurs in coastal salt marshes and adjacent areas from Oregon to central California.

Another HAPLOPAPPUS is *H. venetus* ssp. *vernonioides*, figure 29, from San Francisco south. It is a shrub to about three feet high, somewhat resinous, and very leafy. The heads are without ray-flowers, but have only the yellow tubular ones characteristic of the central part of the heads of so many of the Sunflower Family. One of these is shown to the right in our illustration and exhibits the one-seeded ovary at the base, the hairy modified calyx, the tubular corolla with five lobes representing the petals, and the two-lobed stigma at the summit.

GOLDEN-YARROW (*Eriophyllum lanatum*), figure 30, is a most variable perennial with a somewhat woody base and stems varying in height. The leaves are usually toothed or divided, one-half to three inches long; the flower-heads solitary or in open clusters. They have a hemispheric involucre one-fourth to one-half inch high and bright yellow ray-flowers. Because there are several named varieties of which more than one occurs along the coast, the species in its various

forms may range from British Columbia
to central California. See also page 54.

AMBLYOPAPPUS (*A. pusillus*), figure
31, likewise of the Sunflower Family, has
very small heads without ray-flowers, all
the corollas being minute and tubular.
The plant is a yellow-green, balsamic,
slender annual three to fourteen inches
high and bears slender, entire or somewhat divided leaves. It occurs on beaches,
old dunes, and sea-bluffs from San Luis
Obispo County, California, to Lower
California, and on the Channel Islands.
It flowers from March to June.

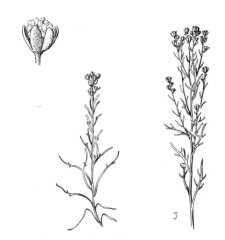

FIGURE 31.   AMBLYOPAPPUS

LASTHENIA (*L. glabrata*), figure 32, is
rather a succulent winter annual found
both in the interior and in salt marshes
of the coast, ranging almost the entire
length of California (see page 54). A
member of the Sunflower Family, its
many small flowers in heads resembling
a solitary flower, it gets to be one to two
feet high, mostly smooth, the entire
leaves one to four inches long. The ray-
flowers are orange-yellow, to almost one-
half inch long. This plant is closely
related to the following and they are
often put in the same genus.

FIGURE 32.   LASTHENIA

BAERIA is the genus name often used
for our common slender-stemmed annual
Goldfields or Sunshine which may cover
miles of our dry interior. Near the coast
is a species *Baeria macrantha*, figure 33,
that lasts more than one year and is more
or less pubescent with ascending hairs.
The leaves are one to four inches long,
narrow, paired as in Lasthenia, but the
ray-flowers are slightly longer than in
that plant. In somewhat different forms
it occurs in grassy places and on coastal
strand from Curry County, Oregon, to
San Luis Obispo County, California.
Flowering is largely from March to
August.

FIGURE 33.   BAERIA

FIGURE 34.   PINCUSHION FLOWER

FIGURE 35.   SNEEZEWEED

FIGURE 36.   JAUMEA

Another species that varies in different parts of its range is the yellow PINCUSHION FLOWER (*Chaenactis glabriuscula*), figure 34, a winter annual four to sixteen inches high; its many yellow florets are arranged in a compact head, but all of them are tubular, the outer somewhat enlarged. The species varies from almost smooth to quite woolly, from large-headed to smaller, from much-divided leaves to less so. It occurs in the interior and coastal regions of much of California, and comes down to the beach in more than one form and at various places from the San Francisco region to San Diego.

SNEEZEWEED (*Helenium Bolanderi*), figure 35, forms clumps one to two feet high with several stout stems from a thick root. It is more or less woolly, especially about the heads, which are solitary at the tips of long naked stems and have a hemispheric center and yellow petallike rays to about one inch long. It is frequent in moist meadows and coastal swamps and on bluffs from Mendocino County, California, to Coos County, Oregon. It is doubtfully distinct from the interior species *H. Bigelovii* and perhaps both should be combined under the name *H. decurrens.*

In salt marshes and wet places on the beach and scattered from northern Lower California to the Puget Sound region and Vancouver Island is JAUMEA (*J. carnosa*), figure 36, a fleshy perennial with creeping branched rhizomes and numerous, mostly simple, more or less prostrate or ascending stems. The heads are generally solitary, with narrow inconspicuous ray-flowers and yellow disk-flowers. The fleshy involucres are characteristic.

Another member of the Sunflower Family and quite an attractive one, but apparently without any generally used common name, is VENEGASIA (*V. carpesioides*), figure 37. It is a leafy, almost or quite hairless perennial, with many stems and thin bright green leaves two to six inches long. The heads have a loose involucre, the outer bracts especially being quite leafy. The bright yellow ray-flowers are almost one inch long and obscurely toothed at the blunt tip. This species is largely a shade plant of rocky and steep places along the coast from Monterey County, California, to northern Lower California.

Quite a large western group of the Sunflower Family is known as the tarweeds because of their glandular or viscid and heavy-scented herbage. Among these are some found along the coast, one being the TARWEED (*Madia madioides*), figure 38. It is a slender-stemmed perennial one to two feet tall and forms a well-developed basal rosette of coarsely hairy, slightly toothed leaves two to four inches long. The heads are few, yellow, showy. This species is found largely in coniferous woods along the coast from Monterey County, California, to Vancouver Island and blooms from July to September.

One of the truly large genera of flowering plants is Senecio, often called Groundsel or Ragwort. It has well over one thousand species, some being trees, others vines and shrubs, but by far the most are herbs. On our coast, growing on dunes and back beaches of San Luis Obispo and Santa Barbara counties, California, is *Senecio Blochmanae*, figure 39, perhaps best called SENECIO. It is an undershrub to about three feet high, with linear-filiform leaves one to three inches long, and flat-topped groups of yellow-

FIGURE 37. VENEGASIA

FIGURE 38. TARWEED

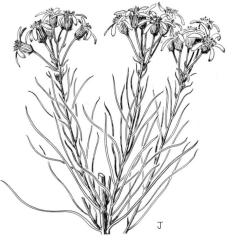

FIGURE 39. SENECIO

FIGURE 40. SENECIO

FIGURE 41. MICROSERIS

FIGURE 42. AGOSERIS

flowered heads about one inch in diameter.

Another quite different SENECIO (*S. Bolanderi*), figure 40, of the immediate coast is a perennial herb with underground rootstocks. Its slender stems are one to two feet tall and produce at the base rounded or somewhat heart-shaped leaves with those of the stem more lobed. The heads are about one-third inch high and one inch across. It is found on the coastal strand and neighboring bluffs from southwestern Oregon to Mendo· cino County, California. Another coastal species is *S. californicus*, a low annual with clasping leaves, occurring from Monterey County, California, south.

One of the most specialized groups in the Sunflower Family has all the florets modified into elongate strap-shaped corollas. Among these is MICROSERIS (*M. Bigelovii*), figure 41, a stemless annual of coastal bluffs and flats from Santa Barbara County, California, to British Columbia. The leaves are two to ten inches long, entire or with toothlike lateral lobes. The heads are solitary at the ends of long stems and each consists of rather numerous, flattened, elongate, petallike florets.

Related to Microseris in having all the florets strap-shaped is Agoseris, mostly called Mountain Dandelion, since it is commonly found in the pine belt of western mountains. But the species along our beaches and their environs had best be designated as AGOSERIS (*A. apargioides*), figure 42, which is found in three named varieties from southwestern Washington to Santa Barbara County, California. It is a perennial, thinly hairy to woolly or almost smooth, with entire to pinnatifid leaves (as in the illustration) and dandelionlike heads an inch or more across.

COLOR PLATES

GIANT HORSETAIL (*Equisetum Telmateia* var. *Braunii*) is shown in plate 1 in its sterile condition in late June. In Equisetum the stems and branches are hollow and jointed, being solid only at the nodes. In the type shown here there is profuse branching, hence the name Horsetail. Early in the spring appear the fertile shoots with terminal spore-producing cones as shown on page 10. Later come the sterile green much-branched stems. The plant grows in swampy and moist places from the seepy banks on the beach to the interior and ranges north to British Columbia.

PLATE 1.   GIANT HORSETAIL

The COLUMBIA LILY (*Lilium columbianum*), plate 2, is not a beach plant, but does occur in the scrub and wooded places near the beach, ranging from Humboldt County, California, to British Columbia and blooming in June and July. The lower leaves are usually in whorls of five to nine, the upper scattered. The perianth-segments are one and one-half to over two inches long, recurved, lemon-yellow to golden to deep red with a yellow center and are usually spotted maroon. Near the shore from Mendocino County, California, south to Marin County is Coast Lily (*Lilium maritimum*), page 61, of more moist places and with leaves usually scattered and flowers horizontal. Western Lily (*Lilium occidentale*) of moist places has only the central leaves whorled.

PLATE 2.   COLUMBIA LILY

The most common IRIS that gets near the coast is *Iris Douglasiana*, plate 3. This species forms heavy sturdy clumps with leaves to three-fourths of an inch wide and flower-stalks to over two feet tall. Flower color varies from pale cream to lavender or deep red-purple. The range is from Santa Barbara County, California, to Oregon and the flowering season from March to May.

PLATE 3.   IRIS

27

PLATE 4.   WILD BUCKWHEAT

PLATE 5.   WILD BUCKWHEAT

PLATE 6.   WILD BUCKWHEAT

For the Buckwheat Family see pages 29, 85–87. One of the most diverse and widespread species of WILD BUCKWHEAT is *Eriogonum latifolium*, plate 4. The typical form is woody and densely leafy at the base, with leaves white-woolly at least on the under surface. The stout leafless flowering stems usually fork and may reach a height of two feet. At their summits appear the small white to rose flowers about one-eighth of an inch long and having three outer and three inner perianth-segments which are petallike. This form is found in sandy places along the coast from San Luis Obispo County, California, to Oregon.

With more glabrous stems, that is, free from hair, and with the leaves scattered along the lower parts of the plant is *Eriogonum latifolium* ssp. *grande*, plate 5, a BUCKWHEAT of the Channel Islands off the southern California coast. The flowers may vary from whitish to quite a deep red. This subspecies grows on bluffs and cliffs near the sea.

Another WILD BUCKWHEAT is *Eriogonum fasciculatum*, often called CALIFORNIA BUCKWHEAT, see plate 6. It is a low spreading shrub with openly branched inflorescence and innumerable white to pinkish flowers that become rusty brown in age. It is common on dry slopes and canyon walls of the immediate coast from Santa Barbara County, California, to northern Baja California and blooms from May to October. Together with its more inland forms it is an important bee plant, being the source of an unusually good quality of honey.

Perhaps the showiest species of Wild Buckwheat is SAINT CATHERINE'S LACE (*Eriogonum giganteum*) of plate 7. Most common on Santa Catalina Island, California, it occurs also in modified form on San Clemente and Santa Barbara islands. The typical form is a coarse rounded branching shrub one to nine feet tall, with a central trunk to several inches thick. The leaf-blades are leathery, more or less white-woolly especially underneath. The great forking cymes of tiny white flowers are very conspicuous and become a beautiful rusty brown in age. Then they make good permanent dry bouquets.

PLATE 7. SAINT CATHERINE'S LACE

In the Four-O'Clock Family are found such plants as Bougainvillea and the garden Four-O'Clock, but the most showy of the native western members are the so-called Sand-Verbenas. Our beach sands have three species, but there are others in the mountains, deserts, and on the plains of North America. Plate 8 shows a yellow-flowered species of SAND-VERBENA (*Abronia latifolia*), common on the coastal strand from Santa Barbara County, California, to British Columbia. The salverform flowers have no petals, the yellow petaloid parts being the modified calyx. The flowers occur in heads subtended by a calyxlike involucre of several bracts. Flowering is from May to October.

PLATE 8. SAND-VERBENA

The next species is *Abronia maritima*, the SAND-VERBENA of plate 9. Like *A. latifolia* it is fleshy, much-branched, prostrate, and sticky-pubescent. The small flowers are dark crimson to red-purple, about one-sixth inch broad, while in the preceding species they are as much as one-third inch across. *Abronia maritima* is found on the lower coastal strand, from Lower California to San Luis Obispo County, California, and blooms from February to October.

PLATE 9. SAND-VERBENA

29

PLATE 10. SAND-VERBENA

The third coastal SAND-VERBENA is *Abronia umbellata* of plate 10, shown as it grows on higher parts of the strand such as dunes and back beaches. It is extremely variable and in one form or another ranges the length of California and far into Lower California. It is less succulent than the other two species and closely resembles *Abronia villosa* of interior valleys and deserts, with which it intergrades. The flowers vary from rose to whitish and appear from spring to autumn.

PLATE 11. ICE PLANT

In the Carpet-Weed Family (see pages 13, 91) the sea beaches have a number of conspicuous plants, one such being ICE PLANT (*Mesembryanthemum crystallinum*) of plate 11. It is an annual, usually with broad alternate leaves, the surface of which is covered with shining, colorless, conspicuous papillae. It is very succulent, prostrate, much-branched, and variously hued. The flowers are somewhat less than an inch in diameter, white to reddish. Ice Plant occurs in sandy or saline places from Monterey County to Lower California. Another annual species with leaves largely alternate, but semiterete instead of flat, is *Mesembryanthemum nodiflorum*.

PLATE 12. SEA-FIG

SEA-FIG is *Mesembryanthemum chilense*, plate 12, with running stems bearing opposite three-sided leaves lacking transparent papillae. The rose-magenta flowers are one to two inches broad, while in *M. edule* (Hottentot-Fig) of page 13, they are yellow, drying pink, and are three to four inches across. Sea-Fig ranges from Oregon to Lower California. Both species are introduced and are commonly used along road cuts and banks to prevent erosion.

30

A common BUTTERCUP in California is *Ranunculus californicus*, plate 13. A perennial with slender roots and with stems to over two feet tall, it is a common plant in fields during the spring months. It ranges from southern Oregon to northern Lower California and comes out to the coast in open places. A more strictly coastal form with prostrate stems is shown on page 13. Another species much like it is *Ranunculus occidentalis*, which usually has five to six petals one to two times as long as broad, while *R. californicus* has largely eight to fifteen, mostly two to three times as long as broad.

PLATE 13. BUTTERCUP

CALIFORNIA POPPY of the Poppy Family is widespread on the west coast, particularly in California, where it occupies many ecological niches, even on the bluffs and beaches along the sea. There the most common species is *Eschscholzia californica*, the typical form occurring from Santa Barbara County to Mendocino County. It is characterized by being heavy-rooted, glaucous, and with smooth broad compact leaves. Plate 14 shows var. *maritima*, a form with very gray roughish puberulent leaves that, under a lens, appear pitted when dry. The stems are prostrate. The range is from Monterey County to Santa Barbara County.

PLATE 14. CALIFORNIA POPPY

Near the Poppy Family is the Fumitory Family, which characteristically has fewer stamens and the petals do not shed so soon. In Dicentra the outer pair of the four petals is saccate or spurred at the base. Such a plant is our western BLEEDING HEART (*Dicentra formosa*), of plate 15, a plant of wooded shaded places, but found sometimes on the immediate coast. It ranges from central California to southwestern Oregon and flowers from March to July.

PLATE 15. BLEEDING HEART

31

PLATE 16.   SEA-ROCKET

PLATE 17.   WALLFLOWER

PLATE 18.   LIVE-FOREVER or DUDLEYA

The Mustard Family (pages 15, 63, 92, 93) with its four sepals and petals and superior ovary is common and has several representatives along the coast. One, fleshy, branched, and glabrous, is an annual group, the SEA-ROCKET (*Cakile maritima*) of plate 16. The pod is fleshy and transversely two-jointed. This introduced plant is found on beach sand from Monterey County to Mendocino County, California. Its leaves are pinnatifid and the petals almost one-half inch long. Another species is *C. edentula* ssp. *californica*, with the leaves merely sinuate-toothed and the petals one-fourth inch long. It is native from San Diego to British Columbia.

Another group in the Mustard Family is WALLFLOWER. In plate 17 is shown *Erysimum suffrutescens*, a species of the coastal strand of central and southern California. It is a much more rampant grower than the ones shown on page 15, has narrow leaves one to three inches long, and bright yellow petals three-fourths inch long. The linear pods are two to three and one-half inches long. It is variable and extends from San Luis Obispo County, California, south.

All along our coast, especially on bluffs overlooking the sea, are succulents of the Stonecrop Family. Unlike the cacti, they are not spiny and do have four or five sepals and petals and quite distinct carpels. LIVE-FOREVER or DUDLEYA has a number of coastal species, one (*Dudleya caespitosa*) being shown in a nonflowering state in plate 18. Its rosette-leaves are two to eight inches long and it bears cymes of bright yellow to red flowers which are an inch or so wide. It is found from Monterey to Los Angeles counties and closely resembles *D. farinosa*, a species ranging from southern Oregon to Los Angeles County and with leaves one to two inches long and pale yellow petals. Other species occur farther south.

Another Live-Forever is *Dudleya pulverulenta* of plate 19. It is a larger plant with a thick stem a foot or so tall, with many rosette-leaves three to ten inches long, and is covered with a white mealy powder throughout. The flowering branches are one to two and one-half feet long and the deep red petals one-half inch or longer. It occurs from San Luis Obispo County, California, to Lower California, blooming from May to July. Related species occur in Lower California.

PLATE 19.   LIVE-FOREVER

A Live-Forever of a different sort is *Dudleya virens*, plate 20, with its corolla segments spreading from near the base instead of only near the tips, and with much narrower leaves. It is found on sea cliffs on Catalina, San Clemente, and Guadalupe islands and has procumbent stems that form large masses with grayish-green leaves and whitish flowers. Surprisingly enough, these plants from a cool insular habitat grow very well in cultivation in interior California, as at Claremont.

PLATE 20.   LIVE-FOREVER

In the same family with Dudleya is Stonecrop, a species of which, *Sedum spathulifolium*, is shown in plate 21. It is a perennial with slender rootstocks and prominent rosettes of leaves which reach a length of about one inch. The erect or decumbent flowering stems grow to about one foot high and bear cymes of yellow to orange or white flowers one-half inch or more across. Along the coast this species and its subspecies grow from California to British Columbia and bear flowers from May to July.

PLATE 21.   STONECROP

33

PLATE 22.   ALUM ROOT

PLATE 23. FUCHSIA-FLOWERED GOOSEBERRY

PLATE 24.   OCEAN SPRAY

For other members of the Saxifrage Family see pages 63, 93, and 94. In plate 22 is represented a species of ALUM ROOT (*Heuchera micrantha*) which grows near the coast from San Luis Obispo County, California, to southern Oregon. It is found on rocky banks and in humus. It has a well-developed caudex, five-to-seven-lobed basal leaves, and stoutish flowering stems one to two feet high. The whitish flowers are minute but borne in great profusion and appear from May to July. Another species of the immediate coast from San Luis Obispo County to Humboldt County, California, is *Heuchera pilosissima* with flowers more rounded at the base, pinkish-white, and with the styles much less prominently exserted.

In this same family is FUCHSIA-FLOW-ERED GOOSEBERRY (*Ribes speciosum*), a red-flowered spiny shrub illustrated in plate 23. It is remarkable among currants and gooseberries in having only four sepals and petals instead of the usual five. The bush is more or less evergreen, three to six feet tall, very spiny, and with glossy deep green leaves. The bright red hanging flowers are striking and come from January to May. The range is on coastal bluffs and in adjacent canyons from Santa Clara County, California, to northern Lower California.

Coming now to several members of the Rose Family (see also pages 64, 95, 96), there is cited first a spiraealike shrub called OCEAN SPRAY or CREAM BUSH (*Holodiscus discolor*), plate 24. It is spreading, four to eighteen feet high, with a number of named forms varying in leaf size and teeth, but in general growing in rocky places such as sea bluffs and canyons on and away from the immediate coast, from British Columbia to southern California. The flowers are whitish and in rather large panicles, appearing from May to August.

34

Another member of the Rose Family is Potentilla, which may have leaves palmately divided, when often called Cinquefoil, or pinnately divided and may then be SILVERWEED as in *Potentilla Egedei* var. *grandis* (plate 25). This plant is a creeping perennial with long runners or stolons and suberect leaves eight to twenty inches long and having seven to thirty-one leaflets white-woolly underneath. The bright yellow flowers are about one inch across and appear from April to August. The variety is found on sandy beaches and in salt marshes from southern California to Alaska.

PLATE 25.　SILVERWEED

Another beach dweller which belongs in the Rose Family is BEACH STRAWBERRY (*Fragaria chiloensis*), plate 26. It too spreads by runners or stolons, but has three leaflets only. These are shining above and silky beneath. The pure white flowers are staminate or pistillate, the two kinds usually being produced on different plants. Living on beaches and adjacent bluffs, this strawberry ranges from central California to Alaska and is found also in Hawaii and South America. The Chilean form is one of the ancestors of domestic strawberries.

PLATE 26.　BEACH STRAWBERRY

A conspicuous member of the Rose Family growing along streams and in moist woods and coming out to the coast from Mendocino County, California, north to Alaska is GOAT'S BEARD (*Aruncus vulgaris*) of plate 27. It is a perennial three to six feet high, with divided leaves and long slender racemes of small white flowers. These are produced from May to July and are followed by small dry follicles or seed-pods one-eighth of an inch long.

PLATE 27.　GOAT'S BEARD

35

PLATE 28. CALIFORNIA BLACKBERRY

PLATE 29. THIMBLEBERRY

PLATE 30. BUSH LUPINE

The Rose Family has many species which produce fleshy fruits like the so-called berries (strawberry, blackberry, raspberry), pome fruits (apple, pear, haw), and stone fruits (plum, cherry, and peach). The CALIFORNIA BLACK-BERRY (*Rubus vitifolius*) of plate 28 is a mound-builder or trailer or partial climber of some size, with bright green rather thin leaves, the leaflets sharp-pointed. The fruit consists of a cluster of minute stone fruits; it is black and less than one-half inch long. It occurs on beaches and adjacent areas from San Luis Obispo County to Mendocino County, California. Much like it is *Rubus ursinus*, with duller, usually less pointed leaflets and larger berries.

A different sort of *Rubus* is THIMBLE-BERRY (*R. parviflorus*), plate 29. It is deciduous, without prickles, and with shreddy bark in age. The leaves are lobed, not divided into separate leaflets, and are four to six inches broad. The fruit is scarlet, hemispheric, about one-half inch in diameter. The species occurs widely, but along the coast may be found near woods and thickets, from Santa Barbara County, California, to Alaska. It flowers from March to August and produces rather flavorless fruit.

One of the largest families of flowering plants is the Pea Family (see pages 16, 37–39, 64, 72, 97) and one of its large western genera is the lupines, which may be annual, perennial, or shrubby. They are found from the seacoast to the highest mountains and to the desert. A conspicuous BUSH LUPINE of the coastal strand is *Lupinus Chamissonis*, plate 30. It ranges from Marin County, California, to Los Angeles County. It is a silky bush two to six feet high, with short leafy branches and six to nine leaflets. The more or less whorled flowers are borne in racemes two to six inches long.

36

Another BUSH LUPINE is *Lupinus arboreus*, differing from the preceding in having the upper petal hairless on the back. The flower is less pointed and more rounded and blunt than in *L. Chamissonis*. It is the most abundant coastal shrubby lupine and has somewhat broader leaflets than does *L. Chamissonis*. It forms great masses, largely at elevations below 100 feet, from Ventura County, California, to Lane County, Oregon. It is remarkable in having at least two color forms, one yellow as in plate 31, the other bluish or whitish or lilac, as in plate 32. Another Bush Lupine and one that may be found also on the California coast is *Lupinus albifrons*, which is not pictured here, but which agrees with *L. Chamissonis* in having hairs on the back of the upper petal. However, it differs in having fine hairs along the upper edge of the distal part of the keel (the boat-shaped pair of petals in which the stamens and pistil lie), while *L. Chamisson* lacks these.

PLATE 31.   BUSH LUPINE

PLATE 32.   BUSH LUPINE

Northward are two perennial coastal species of LUPINE which are not woody: *Lupinus variicolor*, plate 33, and *L. littoralis*, page 71. *Lupinus variicolor* has slender, more or less prostrate stems, somewhat hairy but green foliage, seven to nine leaflets more or less silky beneath. The flowers are about one-half inch long and are yellow, whitish, pink, purple, or blue. It is found from San Luis Obispo County, California, to Humboldt County. More widespread is *L. littoralis*, ranging from northern California to British Columbia and characterized by being more spreading-hairy, with bright yellow roots. Its petioles are one to two inches long, as compared with the one and one-half to four inches of *L. variicolor*.

PLATE 33.   LUPINE

37

PLATE 34.   FURZE or GORSE

Another conspicuous coastal plant with pea-shaped flowers is an introduction from Europe, namely FURZE or GORSE (*Ulex europaeus*), plate 34. It is a wicked thing, very densely branched with thick spinescent branches, simple stiff spinose leaves, and showy yellow flowers. It is naturalized at spots along the coast from southern California, but more abundantly northward, to British Columbia. In some regions, for instance Bandon Beach, Oregon, it is crowding out the native vegetation and forms an almost impenetrable mass. It flowers from February to July.

PLATE 35.   BROOM

A yellow-flowered shrub with pea-shaped flowers, but not spinose, is the introduced BROOM from the Mediterranean region. It is naturalized, often perniciously so, and is crowding out the native shrubbery near the coast from Ventura County, California, north to Washington. In our western books it has been reported as *Cytisus monspessulanus* or French Broom and is illustrated in plate 35. It seems that the correct name may be *Cytisus maderensis*, that is, coming from the island of Madeira. Its stems are obtusely angled or ridged and the pods are hairy all over.

PLATE 36.   SCOTCH BROOM

Another Broom is SCOTCH BROOM (*Cytisus scoparius*) shown in plate 36. It has sharply angled stems that have leafless or almost leafless branches. The pods are hairy along the margins only. It is becoming extensively naturalized from Santa Cruz County, California, northward into Washington and is a native of Europe. In both species the flowers appear in spring and early summer.

38

Among the western American species of the Pea Family is the rather large genus Lotus (see page 16), one species of which, *Lotus scoparius* or DEERWEED, is shown in plate 37. It is by no means confined to the shore, but it is common as a bushy greenish subshrub in many places along the coast of much of California and northern Lower California. The leaflets are mostly three, to about half an inch long. The flowers are in small clusters of one to five, in the leaf-axils, and with the corolla one-third inch long and yellow or tinged with red. They appear from March to August.

Our cultivated Sweet Pea belongs to the genus Lathyrus and on the West Coast we have several species in this genus. One, BEACH PEA (*Lathyrus littoralis*), is illustrated in plate 38. It is a white-silky perennial; the leaflets are four to eight, and the flowers mostly two to six in a group. They vary from white to pink to purple and are half an inch or longer. The species occurs on the coastal strand from Monterery County, California, to British Columbia and bears flowers from April to July.

Another perennial BEACH PEA is *Lathyrus japonicus* var. *glaber* of plate 39. Like most species of Lathyrus it has well developed tendrils. The leaves are green and more or less fleshy. The flowers are two to eight in number, about one inch long, purple or with wings and keel whitish. It grows on the coastal strand from extreme northern California to Alaska and also about the Great Lakes. Flowering is from May to July. Two introduced annual species with only two leaflets are Tangier Pea (*L. tingitanus*), which is not at all hairy, and *L. odoratus*, the cultivated Sweet Pea, which is hairy. Both escape along the coast.

PLATE 37. DEERWEED

PLATE 38. BEACH PEA

PLATE 39. BEACH PEA

39

PLATE 40. MEADOW FOAM

MEADOW FOAM of the False-Mermaid Family, is *Limnanthes Douglasii* shown in plate 40. A low annual herb with alternate, pinnately dissected leaves and solitary three-to-six-merous flowers, Meadow Foam can form great masses in low places which are moist in the spring. It is not primarily a coastal plant, but is occasional, especially as the yellow-flowered form (var. *sulphurea*) on Point Reyes Peninsula, Marin County, California.

PLATE 41. CALIFORNIA-LILAC

In the Buckthorn Family (see pages 73, 98) the most conspicuous western genus is Ceanothus, which falls into two main groups: (1) with thin deciduous stipules (the small appendages at the base of each petiole), alternate leaves, and flowers in terminal panicles; and (2) with stipules having thick corky persistent bases, leaves often opposite, and flowers largely in lateral umbels. The first group is largely spoken of as "California-Lilac," the second as "Buckthorn." A coastal example of CALIFORNIA-LILAC is *Ceanothus griseus*, with green angled branchlets, broad dark green shining leaves, and violet-blue flowers in dense panicles one to two inches long, as shown in plate 41 in a young stage. It occurs in central and northern California. Other similar blue-flowered or white-flowered species are found along the California coast and as far north as Coos and Curry counties, Oregon.

PLATE 42. BUCKTHORN

Plate 42 illustrates an example of a southern California series belonging to group two above, namely BUCKTHORN (*Ceanothus megacarpus*), from along the coast between Santa Barbara and San Diego counties. There are several related species all with thick harsh, more or less toothed to entire leaves. The white flowers come in very early spring.

40

Rather attractive little plants resembling somewhat single-flowered, pink hollyhocks and in the same Mallow Family (pages 65, 66) is CHECKER, a coastal species being *Sidalcea Hendersonii,* plate 43. It grows on steep slopes overlooking the ocean from Tillamook County, Oregon, north to Vancouver Island. The stems are one to two feet high. The lower leaves are not deeply cut, but the upper are more so.

In the Cactus Family (see page 17) we have a succulent spiny family bearing mostly reduced and very ephemeral leaves, but the stems are modified and carry on the photosynthesis usually performed by the green leaves. A well known genus is Opuntia, with jointed flattened or terete stems. An example of the latter type is COAST CHOLLA of plate 44. It is *Opuntia prolifera,* so called because it proliferates, that is, the fruit does not drop off, but forms at its end a new flower bud the next spring. This process may go on for several years and there are formed strings of subglobose, largely sterile fruits set end to end. Found near and on the coast, the species occurs on some of the Channel Islands and on the mainland from Ventura County to nothern Lower California.

Another cactus inhabiting dry bluffs and cliffs in southern California is the GREEN-SPINED CEREUS or *Cereus Emoryi* of plate 45. With prostrate columnar stems from which grow erect branches with fifteen to twenty-five ribs and with yellow spines that darken in age, this species was at one time common on the mainland from Orange County, California, to Lower California, but has now largely been removed by collectors north of the Mexican border. It is found also on San Clemente and Santa Catalina islands, its greenish-yellow flowers appearing in May and June.

PLATE 43.   CHECKER

PLATE 44.   COAST CHOLLA

PLATE 45.   GREEN-SPINED CEREUS

PLATE 46.   CALIFORNIA-FUCHSIA

PLATE 47.   GODETIA

PLATE 48.   GODETIA

In the Evening-Primrose Family (pages 17, 43, 67) are several characteristic western plants, among them CALIFORNIA-FUCHSIA (*Zauschneria cana*), plate 46. A grayish perennial somewhat woody at the base, with narrow fascicled leaves and with brilliant tubular scarlet flowers about one inch long, it commands attention in late summer and early fall. The ovary is inferior, that is, below the flower; sepals and petals are four and the latter two-lobed. It grows in rocky places near the coast from Monterey County, California, southward and also on the Channel Islands.

The so-called GODETIA or FAREWELL-TO-SPRING is a large group of the Evening-Primrose Family belonging to the genus Clarkia (see page 67). They are annual plants usually branched, often with exfoliating epidermis on the lower stems, with a large narrow inferior ovary, and with four entire petals in maritime species. One of these to be met near the coast is *Clarkia amoena* (plate 47), erect or sprawling, one to three feet tall, with leaves to over two inches long, and with the petals an inch or more long, fan-shaped to obovate, pink or lavender to white above, often pink or lavender at the base, and mostly penciled or blotched in the center with bright red. It is found on bluffs and slopes near the sea along the California coast north of San Francisco Bay.

In similar places from Marin and Alameda counties south to Santa Clara and Santa Cruz counties, California, grows *Clarkia rubicunda*, plate 48, a GODETIA with lavender to somewhat pinkish petals, usually with a bright red base but no central blotch. Both species bloom from May to July or August.

The EVENING-PRIMROSE itself, for which the family is named, belongs to the genus Oenothera, one species of which, *Oenothera Hookeri*, is shown in plate 49. It is a tall rather weedy plant, biennial, typically living over the first winter as a rosette of leaves and possessing a fleshy root. *Oenothera Hookeri* is the large-flowered species all through the southwestern states, opening in the late afternoon with flowers an inch or more in diameter. Two or three forms or subspecies, which differ in technical characters of pubescence, length of sepal-tips, et cetera, grow on moist beaches, seeps on the sea bluffs, and stream banks of much of the California coast.

PLATE 49.   EVENING-PRIMROSE

Another member of the genus Oenothera, but one that opens in the morning, may well be called BEACH-PRIMROSE (*Oenothera cheiranthifolia*), plate 50. It usually grows flat on the sandy beach, is perennial, but flowers the first year, has wiry stems, usually silvery or gray foliage, and bright yellow flowers often with central reddish spots. From Point Conception, Santa Barbara County, California, to Coos Bay, Oregon, is a small-flowered form with petals to one-third inch long. From Point Conception south is a woodier form with petals one-half to almost an inch long. See also page 18.

PLATE 50.   BEACH-PRIMROSE

The Carrot Family (pages 18, 44, 73, 99, 100) also has an inferior ovary, that is, below the flower proper. The flowers are small and borne in umbels, all rising from one point instead of up and down an axis. The plants have oil-tubes and are aromatic. Here belongs POISON-HEMLOCK (*Conium maculatum*) of plate 51. It is a tall biennial herb with spotted stems and much compounded leaves. The small white flowers are arranged in many-rayed compound umbels. It is naturalized from Europe and has established itself in low places and in great masses along parts of our Pacific Coast. The flowers appear from April to July.

PLATE 51.   POISON-HEMLOCK

PLATE 52.   COW-PARSNIP

Another member of the Carrot Family is Cow-PARSNIP (*Heracleum lanatum*), plate 52. It is a common plant in the mountains below 9,500 feet, but is also well represented along the coast from Monterey County, California, to Alaska. Perennial, three to eight feet high, somewhat woolly, it has large rounded leaves four to twenty inches broad and more or less lobed. The white flowers are in large flat umbels on peduncles subtended by conspicuous expanded bracts or modified leaves. Flowering is from April to July and produces flattened fruits one-third to one-half inch long.

PLATE 53.   WESTERN LABRADOR-TEA

The Heath Family (pages 18, 45, 46, 100, 101) is woody, usually has the petals united at least at the base and alike, a superior ovary (up in the flower), and is characteristic of cool regions. Belonging to it is WESTERN LABRADOR-TEA (*Ledum glandulosum* ssp. *columbianum*), plate 53, a more or less resinous shrub two to five feet high. The leaves are about one to two inches long, entire, usually revolute on the margins, leathery, entire, and persistent. The small white flowers are in crowded rounded clusters and are followed by capsules about one-fourth inch long. This subspecies is found near the coast from Santa Cruz County, California, to Washington.

PLATE 54.   CALIFORNIA ROSE-BAY

Also in the Heath Family is Rhododendron, a very large genus of Old and New Worlds and important in horticulture. CALIFORNIA ROSE-BAY (*Rhododendron macrophyllum*) has rose-colored flowers as shown in plate 54. It is an evergreen shrub three to twelve feet tall with leathery dark green leaves two to eight inches long and flowers one to one and one-half inches long. It is found on sea bluffs and in shaded woods from Monterey County, California, to British Colunmbia. It blooms from April to July, depending on the latitude.

44

Our other western Rhododendron is WESTERN AZALEA (*Rhododendron occidentale*) of plate 55. It is a deciduous shrub, loosely branched, three to twelve feet tall, with shredding bark and thin, light-green leaves one to three inches long. It grows in moist places, coming out to the coast from Umpqua Valley, Oregon, south to Santa Cruz County, California. Flower color varies from rose to whitish with a central salmon spot to pink with an orange flush. Flowering season is April to June and later in the mountains.

PLATE 55.   WESTERN AZALEA

Another member of the Heath Family is SALAL (*Gaultheria Shallon*), plate 56. It is a spreading subshrub or shrub one to five feet tall, with tough evergreen leaves one to four inches long. The urn-shaped flowers are white to pink, one-third inch long, arranged in racemes or panicles to six inches long. The fruit is a dark purple capsule. Salal grows in brushy or wooded places near the coast from Santa Barbara County, California, to British Columbia and flowers from April to July. Its coriaceous leaves are used in large quantities by florists who in the trade have named them "lemon leaves."

PLATE 56.   SALAL

MANZANITA (pages 18, 100) means little apple and the applelike character of the small reddish fruits is evident in plate 57 of *Arctostaphylos columbiana*. Of the Heath Family, Manzanita is characterized by its small urn-shaped corollas which vary from pink to white, by its stiff evergreen leaves, and often by its reddish bark. There are many species along the coast, varying in habit, height, and hairiness. A well-distributed species is *A. columbiana*, often called HAIRY MANZANITA, much branched, to eight feet tall, with pale gray leaves one to over two inches long. It is found in rocky places from British Columbia to Sonoma County, California.

PLATE 57.   MANZANITA

45

In the Heath Family too is WESTERN-HUCKLEBERRY (*Vaccinium ovatum*), plate 58. It is a stout, much-branched shrub, evergreen, the smooth shining leaves one-half to one and one-half inches long and somewhat toothed on the margins. The bell-shaped flowers are white to pink, about one-fourth inch long, and produce sweet edible black berries one-third inch long. Not a beach plant, it is found in woods, but comes out to the coast in disturbed places, ranging there from British Columbia to central California. It flowers from March to May or June.

THRIFT (*Armeria maritima* var. *californica*), plate 59, belongs to the Leadwort Family (see page 74). It is a tufted perennial, with persistent basal linear leaves and naked stems three to fifteen or more inches high. These bear heads of rose-pink funnel-shaped flowers about one-third inch in length. It grows on coastal bluffs and in sandy places from San Luis Obispo County, California, to British Columbia. The flowers appear from April to August.

The Morning-Glory Family, composed mostly of trailing or climbing plants (pages 47, 102), is a large one of warmer regions. The true Morning-Glory has a number of species near the coast, one of which, BEACH MORNING-GLORY, is *Convolvulus Soldanella*, plate 60. It is a fleshy prostrate perennial from rootstocks, deep-seated in the beach sands. The kidney-shaped, shining, fleshy leaves are one to two inches broad and the rose to purplish corolla one and one-half to two and one-half inches long. The species is common on the strand from San Diego to Washington and occurs also in South America and the Old World, flowering in our area from April to August.

Another MORNING-GLORY is *Convolvulus cyclostegius* of plate 61. It is a perennial vine with trailing to twining stems to several feet in length. The leaf blades are one to two inches long. The white flowers are subtended by a pair of membranous bracts near the calyx and the corolla is from one to almost two inches long and often has purple stripes on the outside. It may turn pinkish in age. The species is found on the beach or more frequently on bluffs and dry slopes near the shore. Its range is from Monterey County, California, to Los Angeles County.

PLATE 61. MORNING-GLORY

A small family of fleshy herbs, parasitic on roots of other plants and itself lacking chlorophyll but turning more or less brown when dry, is the Lennoa Family. A coastal representative is PHOLISMA (*P. arenarium*), plate 62. The part above ground is four to eight inches tall, clumped, with whitish stem aging brown, with bractlike leaves, and with numerous purplish flowers having a white border. It is occasional in sandy places from San Luis Obispo County, California, to San Diego County. The flowers appear for the most part from April to July.

PLATE 62. PHOLISMA

For the Waterleaf Family see pages 48, 75, 103. One of the most characteristic species in California is the so-called WILD-HELIOTROPE (*Phacelia distans*), plate 63. It is an annual, six to thirty inches high, usually branched above, pubescent and somewhat stiff-hairy, with the leaves having toothed to pinnatifid divisions. The bluish flowers are in coiled cymes, are broadly bell-shaped, one-fourth to one-third inch long. It is a common species of much of California and grows in sandy coastal areas from Mendocino County southward. Flowers appear in spring and early summer.

PLATE 63. WILD-HELIOTROPE

47

PLATE 64. PHACELIA

PLATE 65. HELIOTROPE

PLATE 66. FIDDLENECK

An example of another type of PHACELIA is *Phacelia argentea*, plate 64. It is in a perennial group, grayish or hoary with largely appressed hair, leaves entire or few-lobed, and flowers usually pale. *Phacelia argentea* is silvery, with roundish to oval leaves one to two inches long, entire or with two broad basal lobes, and with yellowish-white flowers one-fourth inch broad. It inhabits the coastal strand of southwestern Oregon and northwestern California, but other quite similar species grow as far north as Washington and as far south as southern California, although not usually on the sandy beaches.

The Borage Family (pages 75, 103) resembles the Waterleaf Family in its coiled inflorescences (cymes) and its flower shape, but the ovary forms one-seeded nutlets rather than a several-seeded capsule. Here belongs the genus Heliotropium or HELIOTROPE, a species *H. curassavicum* var. *oculatum* being shown in plate 65. This is a perennial with underground rootstocks that send up scattered shoots four to twenty inches high. It is glabrous and glaucous, has entire succulent leaves, and flowers one-eighth to one-fourth inch wide, white with yellow spots. The plant is characteristic of saline spots, hence it is not surprising to find it growing on the coast, especially in California.

Another member of the Borage Family is FIDDLENECK or Amsinckia, an annual, usually a pungent-bristly herb with a yellow to orange corolla. A species found along the coast is *Amsinckia spectabilis*, plate 66, an inhabitant of sandy places and the borders of salt marshes from Tillamook Bay, Oregon, to Lower California. It blooms from March to June, often growing in masses and usually more or less spreading or prostrate. The orange flowers are one-fourth to one-half inch long.

48

For the Mint Family see page 75. A genus belonging to the family is Stachys or HEDGE-NETTLE, usually found in the West in damp places. A coastal species, especially in seeps and similar places on bluffs and in canyons, is *Stachys bullata,* plate 67. Perennial, with slender stems simple or branched and one to three feet high, it has stiff hairs bent downward on the stem angles. The leaves are one to six inches long; the flowers are in whorls of six, purple, one-half to almost one inch long. It and closely related species extend along much of the Pacific shore.

PLATE 67.   HEDGE-NETTLE

NIGHTSHADE or Solanum has in general wheel-shaped flowers and strong-smelling herbage when crushed. The species shown in plate 68 is *Solanum umbelliferum,* not primarily coastal, but growing on the dunes of Morro Bay, San Luis Obispo County, California, and at other points along the California coast. The bluish-purple flowers are one-half inch or more across and are borne in small open clusters. The fruit is a berry like a small green tomato. Some flowers appear during most of the year. See also page 104.

PLATE 68.   NIGHTSHADE

MONKEY-FLOWER or Mimulus belongs to the Figwort Family (see pages 19, 50, 51, 67, 68, 76). One of the most widespread species is *Mimulus guttatus,* a perennial, almost glabrous herb. Along the coast from Santa Barbara County, California, to Washington and growing largely in wet places, especially seeps in coastal bluffs, is ssp. *litoralis* illustrated in plate 69. The plants are stout, usually one to two and one-half feet tall and the flowers are bright yellow with red spots. They are for the most part one and one-half to almost two inches long.

PLATE 69.   MONKEY-FLOWER

49

PLATE 70.   BUSH MONKEY-FLOWER

PLATE 71.   FOXGLOVE

PLATE 72.   PAINT-BRUSH

Another species of Mimulus is the BUSH MONKEY-FLOWER (*Mimulus aurantiacus*), plate 70. It with other shrubby species is often placed in a separate genus Diplacus. This species is a profusely branched shrub, commonly two to four or five feet tall, glandular, and more or less viscid. The leaves have the veins of the upper surface impressed and the margins are often revolute. The flowers are deep-orange to yellow-orange, one and one-half to almost two inches long. It grows in rocky places often on the immediate coast, from western Oregon to south-central California and blooms from March to August.

Another member of the Figwort Family is a species introduced from Europe and now well established in more or less shaded places near the coast from Santa Barbara County, California, to British Columbia. It is FOXGLOVE (*Digitalis purpurea*), of plate 71, a stoutish biennial two to six feet high, with large lower leaves and terminal racemes of showy purple to whitish flowers. The corolla is declined, somewhat inflated distally, and about two inches long. Flowering is from May to September.

PAINT-BRUSH or Castilleja has many western species and several along the immediate coast. It, too, belongs to the Figwort Family. It has a long narrow corolla often with a very short lower saccate lip. Most of the color is in the bracts subtending the flowers. *Castilleja latifolia*, plate 72, has leaves less than three times as long as wide, blunt and sessile. The corolla is about one inch long. The species is found in sandy places along the coast of northern California, while closely related ones appear farther south and as far north as Washington.

50

Another somewhat woody PAINT-BRUSH is *Castilleja foliolosa,* plate 73. It is bushy, white-woolly throughout, one or two feet tall, with narrow leaves, the uppermost with one or two pairs of lobes. The corolla is scarcely one inch long. This species is found in dry rocky places in the Coast Ranges from Humboldt County, California, to northern Lower California and, somewhat, even farther inland. A plant somewhat similar in its woolliness is *Castilleja hololeuca,* but it is insular and has the upper corolla-lip yellowish with pale thin margins instead of greenish with reddish margins.

PLATE 73.   PAINT-BRUSH

The genus Plantago, with its own family, is quite cosmopolitan. In America we have a number of bad weeds belonging in Plantago and introduced from Europe, but we have also a good many native species, one of which, the seaside PLANTAIN, *Plantago maritima* ssp. *juncoides,* is seen in plate 74. This is a small perennial with many strongly ascending narrow leaves and a few somewhat longer scapes (flowering leafless stems) bearing spikes of small greenish, four-petaled flowers. It grows in salt marsh and on the strand and, with its variety *californica,* extends from Santa Barbara County, California, to Alaska. It flowers from May to September.

PLATE 74.   PLANTAIN

Everyone knows ELDERBERRY, especially the blue-berried forms, but in plate 75 is shown a red-berried species, *Sambucus callicarpa.* This is a shrub six to eighteen feet high, with leaves having five to seven leaflets. The inflorescence is two to four inches across and has many small whitish five-lobed flowers. The bright scarlet fruits are about one-sixth inch in diameter. Found on flats and coastal bluffs, the species ranges from British Columbia to central California.

PLATE 75.   ELDERBERRY

51

PLATE 76. TWINBERRY

PLATE 77. BIG ROOT

PLATE 78. BUSH-SUNFLOWER

Both ELDERBERRY of the preceding page and TWINBERRY belong to the Honeysuckle Family, a woody family with united petals often forming two-lipped corollas. TWINBERRY (*Lonicera involucrata*), plate 76, is an upright shrub and its coastal form with which we are concerned (var. *Ledebourii*) is three to ten or more feet tall. The flowers are in pairs arising from the leaf-axils, have a yellowish corolla often tinged red, one-half inch or longer. They are subtended by large broad bracts which may become purplish or reddish and almost inclose the black fruit. Flowering from March to April, it ranges from Santa Barbara County, California, to Alaska.

In the Gourd Family, which is usually climbing, tendril-bearing, and with hard-rinded fruits, we have on the West Coast the genus Marah with several species. One, which may clamber over slopes and vegetation along the immediate coast, is *Marah oreganus*, BIG ROOT or WILD-CUCUMBER of plate 77. With an immense underground perennial tuber and with stems three to twenty feet long, it produces many shallowly lobed leaves. The whitish flowers seen in the illustration are staminate, the pistillate being fewer and producing the green fruit two to three inches long and with or without spines. It grows from British Columbia to central California.

For the Sunflower Family see pages 20–24, 53–58, 68, 76, and 104. Plate 78 shows BUSH-SUNFLOWER (*Encelia californica*), a low rounded subshrub with green leaves one to two and one-half inches long and showy sunflowerlike heads two to three inches across. It inhabits coastal bluffs and inland canyons from Santa Barbara County, California, to Lower California, blooming from February to June. The rays are yellow, the disk purplish-brown.

52

Likewise in the Sunflower Family is the genus Coreopsis of which several species are cultivated. Along the California coast are two perennials of some size and with stout stems and fleshy leaves pinnately divided into linear segments. The large heads are yellow and two to three inches across. Of these two species of COREOPSIS, *C. maritima*, plate 79, grows on coastal bluffs and dunes of San Diego County and of northern Lower California and the adjacent islands. An herbaceous perennial, it has many stems one to almost three feet long, leaves two to ten inches long, and few heads on stout naked peduncles six to twenty inches long.

The other COREOPSIS, *C. gigantea*, of plate 80, is also a spring bloomer. The plant is shrubby, with usually one trunk three to several feet high, simple or branched, and bearing in the rainy season terminal tufts of leaves two to ten inches long and clusters of yellow-flowered heads. In the dry season flowers and leaves are shed and all that remains is a dead-looking stick. It ranges from San Luis Obispo County to Los Angeles County and is also on most of the Channel Islands.

Another member of the Sunflower Family, but with smallish heads of inconspicuous florets, is SILVER BEACHWEED (*Franseria Chamissonis* and its ssp. *bipinnatisecta*), plate 81. A silvery-canescent perennial with radiating procumbent stems, it forms loose mats on the sand. The leaves are simple, toothed to lobed, or incised to pinnatifid (especially in the less silvery subspecies). The stems bear terminal spikes of heads of staminate florets and below these appear the solitary pistillate heads that are spiny and become burlike. The range is from British Columbia to Lower California.

PLATE 79. COREOPSIS

PLATE 80. COREOPSIS

PLATE 81. SILVER BEACHWEED

PLATE 82. LASTHENIA

PLATE 83. GOLDEN-YARROW

PLATE 84. ERIOPHYLLUM

In the Sunflower Family is a small succulent winter annual with pairs of leaves which are narrow and somewhat sheathing at the base. The heads are yellow-flowered, with a hemispheric involucre of united bracts and conspicuous rays or ligules one-fourth to almost one-half inch long. This LASTHENIA (*L. glabrata*), plate 82 (see also page 21), is colonial, forming great masses of yellow in heavy, more or less saline or alkaline soils, practically the length of California. Flowers come from March to May.

With heads resembling those of the foregoing species in their single row of upright involucral bracts is the GOLDEN-YARROW (*Eriophyllum staechadifolium*), plate 83 (see page 20). It is shrubby, much branched, one to five feet high, woolly especially when young, with rather narrow leaves which may be .entire, few-lobed, or, in the var. *artemisifolium*, pinnatifid. They are permanently white-woolly beneath. The heads are numerous, in rather dense corymbs; the involucre is about one-fourth inch high, the yellow rays one-sixth inch long. The species and its variety grow along the coast, on beaches and on bluffs, from Coos County, Oregon, to Santa Barbara County, California, and bloom from April to September.

Quite a different type of ERIOPHYLLUM is *E. multicaule*, plate 84. It is a loosely woolly annual, one to six inches high, diffusely branched. The leaves are almost wedge-shaped, toothed or lobed at the apex, about one-third inch long. The boat-shaped involucral bracts loosely enclose the outer achenes (seeds). The yellow ray-flowers are very short. This plant grows in old sandy fields and on dunes from Monterey County, California, to San Diego County and flowers in the spring.

54

The three species treated on this page are also of the Sunflower Family and in them what may seem to be one flower is, on close inspection, a head of many small flowers or florets. Here belongs GUM PLANT or Grindelia, a perennial with a gummy or resinous exudation, especially about the heads. *Grindelia stricta* ssp. *venulosa* of plate 85 is a plant of coastal marshes and seaside bluffs, ranging from Coos County, Oregon, to Monterey County, California. It is a more or less procumbent perennial to three feet across, with yellowish or whitish stems and serrulate resinous-punctate leaves. The heads are from one to almost two inches across and with recurved involucral tips.

PLATE 85. GUM PLANT

Goldenrod is a group of perennial herbs with leafy, usually simple stems and alternate leaves. The species shown in plate 86, WESTERN GOLDENROD (*Solidago occidentalis*) is stout, three to six feet tall, glabrous, and bears lance-linear leaves two to four inches long. The flowers are in small heads with fifteen to twenty-five short ray-flowers and seven to fourteen disk-flowers. Not primarily a shore inhabitant, it does reach the coast in moist places and ranges from British Columbia to Lower California and Texas. It blooms from July to November.

PLATE 86. WESTERN GOLDENROD

SEASIDE DAISY (*Erigeron glaucus*), of plate 87, is a low perennial with stems from four to sixteen inches high. The leaves are entire or somewhat toothed, mostly basal, three to six inches long. The heads terminate long branches and have very numerous (perhaps to 100) pale violet to lavender rays about half an inch long. Common on coastal bluffs and beaches from Clatsop County, Oregon, to central California, it blooms from April to August.

PLATE 87. SEASIDE DAISY

55

PLATE 88.   WILD TANSY

PLATE 89.   BEACH SAGEWORT

PLATE 90.   YARROW

A typical member of the Sunflower Family is WILD TANSY (*Tanacetum Douglasii*), plate 88. A perennial from stout underground rhizomes, it becomes one to two feet high and has pinnately dissected leaves to about eight inches long. The heads have very short but more or less evident rays and the plant is more or less hairy, but not whitish. It is found on the coastal strand from Humboldt County, California, to British Columbia. A similar species, *T. camphoratum*, white-woolly especially on younger parts and with its rays not at all evident, is found about San Francisco Bay.

There is a series of white-woolly, not very woody species of Artemisia on sandy shores of large lakes and seas of much of the northern hemisphere. Such a species is BEACH SAGEWORT (*Artemisia pycnocephala*) of plate 89, more or less woody at the base, mostly one to two feet tall, densely leafy and whitish or grayish silky-woolly, with leaves dissected into linear lobes. The many small heads are borne in dense panicles. This is a plant of the beaches from Monterey County, California, to southern Oregon. Farther north it is replaced by similar forms of *A. campestris* with more appressed and silky pubescence.

YARROW is an aromatic perennial herb with pinnately dissected leaves and many small heads of flowers arranged in flat-topped clusters. Along the coast we have *Achillea borealis* ssp. *arenicola*, shown in plate 90. The stems are one to two feet tall and the leaf-segments very numerous and thickish. An inhabitant of the coastal strand from Santa Barbara County to Del Norte County, California, it blooms in June and July. Farther north other species of the same general appearance may draw near the coast in their distribution.

The thistles are of course also in the Sunflower Family with very numerous minute florets compacted into a head surrounded by an involucre of bracts. A well distributed coastal THISTLE is *Cirsium brevistylum* of plate 91. The one shown is growing on an almost vertical bank back of the beach near Charleston, Oregon. The range is from British Columbia to southern California. It is a short-lived perennial three to five feet high, more or less crisp-arachnoid, leafy to the top, the leaves more or less loosely woolly beneath and to six inches long. The flowers are dull purple-red.

PLATE 91.   THISTLE

Next come two species of Chrysanthemum, a large genus of the northern hemisphere, having the involucre made up of two to four series of overlapping bracts. Many species are quite aromatic. The CORN CHRYSANTHEMUM (*C. segetum*) of plate 92, is a yellow-flowered species naturalized along the coast and in fields of coastal central and northern California. It is an annual, one to two feet high, much branched, with leaves cut or pinnatifid, usually with a clasping base and with heads one to two inches across. It is a native of the Mediterranean region.

PLATE 92.   CORN CHRYSANTHEMUM

The other species is white, perennial, with solitary heads one to two inches across. It too is a native of the Old World and for a long time has been naturalized in the eastern United States, more recently and increasingly it has established itself in the Pacific states, especially northward. It is OX-EYE DAISY (*Chrysanthemum Leucanthemum*) shown in plate 93. The leaves are for the most part not divided and the plant is almost hairless. Flowering is largely in late spring and early summer.

PLATE 93.   OX-EYE DAISY

57

PLATE 94.   PEARLY EVERLASTING

PLATE 95.   CAT'S EAR

PLATE 96.   MALACOTHRIX

Another representative of the Sunflower Family is not primarily a beach plant, but in the north comes down to the very edge of the sand. It is PEARLY EVERLASTING (*Anaphalis margaritacea*) of plate 94. A white-woolly perennial with slender running rootstocks, it forms patches in which the erect stems are commonly one to two and one-half feet high. The leaves are narrow, sessile, one to four inches long. The bracts of the involucre are the conspicuous feature of the flower-head, being pearly white. Ranging in the coastal regions from Monterey County, California, to Alaska, it is farther south in the mountains.

CAT'S EAR (*Hypochoeris radicata*) of plate 95 is a showy dandelionlike plant in much of California and northward and transcontinental, having been introduced from the Old World. It is a perennial, with hispid pinnatifid leaves two to six inches long and stems one to three feet high. The bright yellow heads, an inch or more across, have all the florets strap-shaped or ligulate. It blooms much of the summer, but in June is very conspicuous in open places in woods and on sea bluffs as well as along roadsides.

Another plant with florets transformed to elongate flattened ligules is Malacothrix (see page 104). MALACOTHRIX (*M. incana*) of plate 96 is a stout-rooted perennial with a woody crown. The herbage is white-woolly and the several stems grow to about one foot in length. The leaves are two to four inches long, subentire to pinnately lobed. The flower-heads are mostly in one's or two's, have much imbricate involucres, and yellow heads about one inch across. The species is one of the coastal strand, from San Luis Obispo County, California, to Ventura County and on Santa Rosa and San Miguel islands. It flowers most of the year.

58

FLOWERS ROSE TO PURPLISH-RED

Section Three

The Coast Lily (*Lilium maritimum*),
figure 43, is a member of the Lily Family
to which belong many of our common
garden plants (see page 27). This lily
has an elongate rhizomatous bulb and
stems one to four or more feet high. The
leaves are usually scattered, not whorled,
dark green, one to five inches long. Flowers are horizontal, bell-shaped, dark red,
maroon-spotted, about one and one-half
inches long. It grows in sandy soil or on
raised hummocks in bogs or in brush and
woods, and ranges from Marin to Mendocino counties, California. The Western
Lily (*Lilium occidentale*) also occurs
near the coast, but has nodding, dark
orange flowers usually spotted maroon.

FIGURE 43.   COAST LILY

Stream Orchis (*Epipactis gigantea*),
figure 44, of the Orchid Family (see page
84) has the characteristic inferior twisted
ovary of that family. It has a creeping
rootstock with fibrous roots, simple leafy
stems to almost three feet tall, and flowers with greenish deeply concave sepals,
but purplish or reddish petals. The lip
is strongly veined and marked with purple or red. Growing along moist stream
banks at low elevations, it is sometimes
found right on the shore; its total range
is from Lower California to British Columbia, South Dakota, and Texas.

FIGURE 44.   STREAM ORCHIS

Figure 45 is of Wild-Ginger (*Asarum
caudatum*) of the Birthwort Family to
which the mostly tropical Dutchman's
Pipe belongs. Our plant is low with slender spicy-smelling rootstocks and crushed
foliage. The leaves are evergreen, one to
four inches long and the brownish-purple
flowers appear from May to July. The
plant grows in deep shade and may occur
in woods right down to the edge of the
beach. It ranges from central California
to British Columbia.

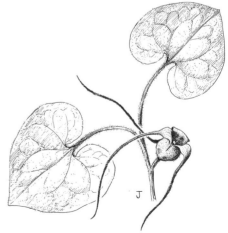

FIGURE 45.   WILD-GINGER

FIGURE 46. AUSTRALIAN SALTBUSH

FIGURE 47. CAMPION or CATCHFLY

FIGURE 48. CALANDRINIA

The AUSTRALIAN SALTBUSH (*Atriplex semibaccata*), figure 46, belongs to the Goosefoot or Pigweed Family known for its small inconspicuous flowers that lack petals (pages 88-90). Here come such plants as Beet, Spinach, and Russian-Thistle. Atriplex has the seed borne between a pair of bracts, one of which is shown in the upper righthand corner of the illustration. The species of Saltbush or Atriplex are largely identified by the shape of these bracts (pages 88, 89). The Australian Saltbush is a prostrate perennial with scurfy grayish hairs and reddish fleshy fruiting bracts. It has become established in saline places in the interior and along the coast from central California south.

A member of the Pink Family (pages 90, 91) is CAMPION or CATCHFLY of which group is here shown *Silene Scouleri* ssp. *grandis,* figure 47. A perennial with one to few stems from a heavy crown, it is as much as two feet tall, with leaves one to six inches long. The ten-nerved calyx is glandular, about one-half inch long, and the divided petals are rose to greenish-white. It grows on bluffs along the coast from San Mateo County, California, to British Columbia.

The Portulaca Family is usually fleshy and has two sepals (page 92). It is represented in figure 48 by CALANDRINIA (*C. maritima*). This is a glaucous annual with more or less spreading stems to ten inches long. The leaves are largely near the base of the plant. The sepals are rounded and dark-veined, the petals red and about one-fourth inch long. It grows in sandy places and on sea bluffs from Santa Barbara County, California, to northern Lower California and bears flowers from March to May.

ROCK-CRESS (*Arabis blepharophylla*),
figure 49, is of the Mustard Family
(pages 15, 32, 92, 93). It is a low perennial with one to few simple stems, the
lower leaves in rosettes and coarsely
hairy. The sepals are oblong, purplish,
about one-third inch long, while the rose-
purple petals measure one-half inch or
longer. The erect smooth pods are an inch
or more. It is a pretty little species of
rocky places along the coast from Santa
Cruz County, California, to Sonoma
County, flowering from February to
April.

In the Saxifrage Family (pages 34,
93-95) which has a tube at the base of
the flower and with four or five sepals
and petals, we find some common plants
like Currant, Gooseberry, Mock-Orange,
Alum Root, and Hydrangea. In figure 50
is depicted TOLMIEA (*T. Menziesii*), a
perennial herb with scaly rootstocks and
chiefly basal leaves. The flowers are in
elongate racemes and bear five unequal
sepals (three long and two short) and
five elongate threadlike petals. The flower tube has some purple and the petals
are brownish. The species reaches the
coast in moist cool places from northern
California to Alaska and flowers in May
and June.

Related to Tolmiea is TELLIMA (*T.
grandiflora*), figure 51, another plant with
horizontal rootstock and both basal and
cauline leaves. The plant is quite hairy
and one to two and one-half feet high.
The bell-shaped flower is about one-sixth
inch long, the petals about one-fourth
inch and are whitish at first, later red.
This species too is not primarily a shore
plant, but it reaches the coast in wooded
places and ranges from central California
to Alaska. It flowers from April to June.

WILD ROSE (*Rosa nutkana*), figure 52,

FIGURE 49.   ROCK-CRESS

FIGURE 50.   TOLMIEA

FIGURE 51.   TELLIMA

FIGURE 52.   WILD ROSE

FIGURE 53.   CLOVER

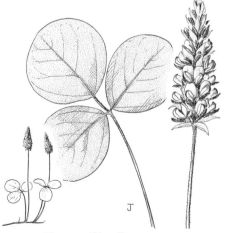

FIGURE 54.   PSORALEA

is not primarily a shore plant, but in the north it reaches the coast in wooded areas. It is a stout-stemmed plant, mostly armed with straightish heavy prickles and growing to a height of three to six feet. The fragrant pink flowers are three or more inches broad and the rounded fruit over one-half inch in diameter. It is found on damp flats and slopes and ranges from northern California to Alaska and the Rocky Mountains. In southern California one may find the common California Wild Rose reaching the coast and may distinguish it by its entire, not toothed sepals and by its recurved prickles. See pages 34–36, 95, and 96 for other members of the family.

The clovers belong to the Pea Family (see pages 16, 36–39, 72, 97) and are characterized not only by the pea-shaped flowers, compound leaves usually with three leaflets, but also by the short, generally one–two-seeded pods. A CLOVER that can be found on the immediate coast is *Trifolium Wormskioldii*, figure 53, named for its Danish discoverer. It is perennial from creeping rootstocks and has branching rather coarse stems. The flowers are about one-half inch long, whitish to purplish-red. It occurs in wet places and is quite common in seeps on rocky bluffs from central California to British Columbia.

Another member of the Pea Family is PSORALEA (*P. orbicularis*), figure 54, of moist places in much of California and appearing along the shore in appropriate spots. It has prostrate stems with long petioles to twenty inches high, leaflets one to three inches long and flower-stalks one to two feet tall. The flowers are reddish-purple, over half an inch long, and appear in early summer. The heavy-scented foliage has dots or glands.

The Crowberry Family is small and has low, heathlike, evergreen shrubs with slender freely branched stems and rigid narrow leaves. CROWBERRY (*Empetrum nigrum*), figure 55, has the stems prostrate or spreading, to about one foot long. The leaves are one-fourth inch long and thickened (upper part of illustration). The minute purplish flowers are solitary in the axils. Sepals and petals are usually three. The black or red berry has several nutlets. Crowberry occurs in dense beds in rocky places on sea bluffs from extreme northern California to Alaska, blooming in the spring.

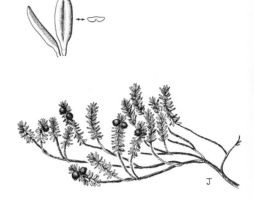

FIGURE 55.   CROWBERRY

In the Sumac Family, a woody group often with poisonous or acrid sap, we find Poison-Oak, Cashew, Mango, and Pistacia. Our western LEMONADEBERRY (*Rhus integrifolia*), figure 56, is a rounded or, near the sea, often flattened shrub three to nine feet tall, with reddish stout twigs and flat coriaceous entire or toothed leaves one to two inches long. The flowers are in dense clusters, more or less pinkish or rose and quite small. They produce flattened, viscid, acid fruits almost half an inch in diameter. It is common on sea bluffs and in coastal canyons from Santa Barbara County, California, south.

FIGURE 56.   LEMONADEBERRY

The Mallow Family usually has the stamens forming a more or less complete tube around the several styles. One of the common genera is CHECKER (*Sidalcea malviflora*), figure 57, a perennial with widely spreading rootstocks and stems one-half to two feet tall. The basal leaves are mostly entire, the cauline deeply lobed. The rose or pink flowers are borne in elongate racemes and are one to two inches across. It occurs in grassy, often damp places, from the coast inland and from southern Oregon to the Mexican

FIGURE 57.   CHECKER

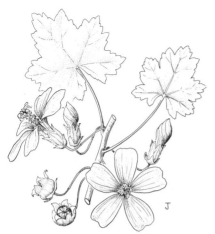

FIGURE 58.   MALVA ROSA

FIGURE 59.   LOOSESTRIFE

FIGURE 60.   FRANKENIA

border. Flowers appear in spring and summer. See page 41.

Another member of the Mallow Family is MALVA ROSA (*Lavatera assurgentiflora*), figure 58, apparently originally native of the islands off the California coast, but planted on the mainland and now abundantly escaped at least in the southern part. It is a shrub or small tree, three to twelve feet high, with large leaves two to six inches wide. The petals are one to two inches long, rose with darker veins. The illustration shows the stamens of several lengths forming a tube around the central styles. Flowers are from March to November.

Figure 59 is LOOSESTRIFE (*Lythrum californicum*) of the Loosestrife Family, to which belongs Crepe-Myrtle, a tree commonly grown in warm regions. Erect and somewhat woody at the base, this Loosestrife grows to over four feet high and has pale green narrow leaves to about one inch long. The flowers have purple petals one-fourth inch long. It is found in moist places near the coast and away from it, from central California southward. It bears flowers from April to October.

The Frankenia Family is small and is represented on the West Coast by two species, one of which, FRANKENIA (*F. grandifolia*), is illustrated in figure 60. It is bushy, somewhat woody at the base, to one foot high, the lower leaves being united in pairs by a membranous base. The scattered flowers are small, pinkish, and the seed pod is linear. Frankenia is found in salt marshes and on moist beaches from Marin and Solano counties, California, to Lower California. It flowers from June to October.

In the Evening-Primrose Family a conspicuous western group is FAREWELL-TO-SPRING or GODETIA (*Clarkia Davyi*), figure 61, with its four-petaled flowers and inferior ovary. This species is prostrate or nearly so, with simple or branched stems and rather crowded leaves less than one inch long. It is found in mostly sandy places along the coast from Del Norte to San Mateo counties, California, and flowers in June and July. See also pages 17, 42, 43.

Another CLARKIA is *Clarkia unguiculata* (*C. elegans* in the older books), figure 62. It is erect, one to three feet high, with leaves to over two inches long. The petals are narrowed basally into a slender claw; they vary from lavender-pink to salmon or purplish or dark red-purple. The species is common on dry, often on shaded slopes, but occurs at the coast even on back beaches. It ranges the entire length of California and blooms in May and June. It was early taken to Europe where double and many other horticultural forms were developed and now appear in gardens.

In the Figwort Family (see pages 19, 49–51, 68, 76) is the FIGWORT itself (*Scrophularia californica*), figure 63, a coarse perennial, three to five feet tall and with large, more or less triangular leaves and numerous flowers in large terminal panicles. The corolla is red-brown to maroon and about one-half inch long. The species is found near the coast, often in brushy and damp places from Los Angeles County, California, to British Columbia. It blooms from February to July. Other forms occur away from the coast.

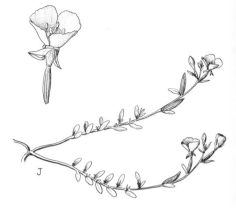

FIGURE 61.

GODETIA or FAREWELL-TO-SPRING

FIGURE 62.   CLARKIA

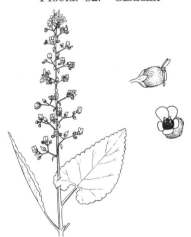

FIGURE 63.   FIGWORT

FIGURE 64.   BIRD'S-BEAK

FIGURE 65.   MARSH-FLEABANE

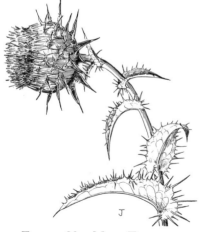

FIGURE 66.   MILK THISTLE

In the same Figwort Family is BIRD's-BEAK (*Cordylanthus maritimus*), figure 64, a branched annual, often with the stems more or less prostrate or nearly so. It is hairy, some of the hairs being gland-tipped. The leaves and bracts are bright glaucous-green and to one inch long. The tubular calyx is one-half to almost one inch long, with terminal short sharp teeth. The corolla is tubular and more or less purplish. The plant occurs in salt marshes along the coast from southern Oregon to northern Lower California and blooms from May to October.

Figure 65 is of MARSH-FLEABANE (*Pluchea purpurascens*) of the Sunflower Family (see pages 20–24, 52–58, 76, 104). It is a glandular-puberulent herb one to three or more feet tall, with leaves two to four inches long and with large terminal flattish clusters of small heads of purplish tubular florets. Ray-flowers are wanting. This plant is found in marshy places along the coast and away from it, from San Francisco Bay south, also on the Atlantic Coast. It blooms from July to November.

MILK THISTLE (*Silybum Marianum*), figure 66, is also in the Sunflower Family. It was introduced from Europe years ago and has become a common pasture weed, but has established itself also on back beaches and dunes. It is erect, branched, three to five feet tall and bears large wavy or crisped leaves mottled with white blotches. The purple florets are in hemispherical heads with spine-tipped involucres as in a true thistle. Flowers bloom from May to July and produce numerous seeds that are carried hither and yon by the downy tuft on each seed.

FLOWERS BLUE TO VIOLET

Section Four

Among the most attractive members of
the Lily Family (see pages 27, 61, 82,
83) are the MARIPOSA-LILIES, one of
which, *Calochortus uniflorus*, is shown in
figure 67. Its flower is built on the plan
of three instead of four and five like that
of broad-leaved plants. It has an underground bulb, few linear leaves, and one
to five lilac flowers, often with a purple
spot above the gland on each petal. The
petals are to about one inch long. It
grows in low, wet, often alkaline places
from Monterey County, California, to
southwestern Oregon and blooms from
April to June.

FIGURE 67.   MARIPOSA-LILY

By way of contrast with the above lily
is LARKSPUR (*Delphinium Menziesii*),
figure 68, with broad net-veined leaves
and flowers on the plan of five. It has
a small shallow underground cluster of
tubers, stems one to two feet high, and
soft spreading white hairs. The leaves
are usually one to two inches wide. The
flowers are few, with deep blue sepals
one-half inch long and smaller paler
petals. It is found in open places on
bluffs above the ocean from Mendocino
County, California, to British Columbia
and may bear flowers from March to
May.

FIGURE 68.   LARKSPUR

Pacific States' LUPINES are a diverse
group with perhaps one hundred species
represented, some annual, some matted,
some perennial, some shrubby. Along the
immediate coast an interesting species is
Lupinus littoralis, figure 69, a slender-
stemmed perennial mostly prostrate or
nearly so. The pea-shaped flowers are
about one-half inch long, blue or lilac,
the roots are yellow, and the leaf-petioles
one to two inches long. It is found along
the immediate coast from northern California to British Columbia. Closely related to it is another species, *Lupinus*

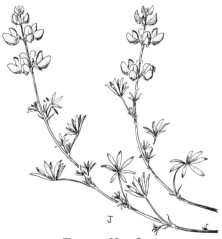

FIGURE 69.   LUPINE

FIGURE 70.   LUPINE

FIGURE 71.   LUPINE

FIGURE 72.   VETCH

variicolor, also coastal, the roots not yellow and the petioles two to four inches long. It ranges from Humboldt County, California, south to San Luis Obispo County. See pages 36, 37, 72.

Another LUPINE is an erect annual with petals rich blue except for the white spot on the banner (upper petal). It is *Lupinus nanus* and is portrayed in figure 70. It is mostly a plant of the interior, covering great areas of grassy fields with its lovely blue. But it comes down to the coast and the bluffs and the beaches, where is may be sought from Mendocino County, California, to Los Angeles County, particularly in April and May. It attains a height of one-half to one and one-half feet and the whorled flowers are about one-half inch long.

Still another LUPINE is *Lupinus rivularis*, figure 71, of wet or sandy places along the immediate coast from Mendocino County, California, to British Columbia. It is a more or less hairy perennial herb, one to three or more feet tall, with petioles one to two inches long and leaflets one to one and one-half inches. The flowers are blue, purplish to almost reddish, somewhat whorled, one-half inch long, and numerous in long racemes. In many ways the most conspicuous beach lupines are the shrubby species; see pages 36, 37.

An introduction originally from Europe is VETCH (*Vicia gigantea*), figure 72. A vinelike herb, it is a stout-stemmed perennial, somewhat pubescent, two to three feet high, with many leaflets on the tendril-bearing leaves. The flowers are reddish-purple, half an inch long, and numerous, in one-sided racemes. It is an inhabitant of moist places near the coast, but is rarely actually out on the open beach, from San Luis Obispo County, California, north to Alaska.

The so-called CALIFORNIA-LILAC or Ceanothus is one of the diverse and highly developed genera of the Pacific States, especially in California. In figure 73 is shown *Ceanothus dentatus,* a densely branched evergreen shrub attaining a height of two to five feet. The branchlets are hairy and the small, toothed leaves crowded, fascicled, and rather narrow. The small deep blue flowers are arranged in clusters to two inches long. It occurs in sandy and gravelly places near the coast from Santa Cruz County to San Luis Obispo County. Other species extend much farther north.

FIGURE 73.   CALIFORNIA-LILAC

Our western VIOLETS have several colors (see page 17) and a bluish one of scrubby bluffs and banks along the coast is *Viola palustris,* figure 74. It is a smooth creeping perennial with a slender rootstock and somewhat roundish leaves two to four inches high and one to two inches broad. The flowers are lilac to almost white with some darker veining, the lateral petals being somewhat bearded. It ranges from Mendocino County, California, to Alaska and occurs also on the Atlantic Coast and in the Old World. It blooms largely from May to July.

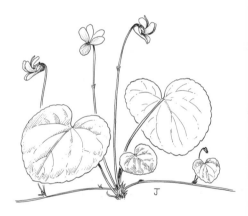

FIGURE 74.   VIOLET

The Carrot Family (pages 18, 43, 44, 99, 100) is large and usually characterized by its aromatic nature due to the contained oils. Some members of the family do not have the well-developed umbels we usually associate with the group. Such an example is ERYNGIUM (*E. armatum*), figure 75, of moist places that dry with the advancing season. Eryngium is to be found near the coast from Santa Barbara County, California, to Humboldt County. It is a spiny plant with stems to over one foot long. The flowers are minute and arranged in tight heads surrounded by bluish or yellowish bracts.

FIGURE 75.   ERYNGIUM

FIGURE 76.
SEA-LAVENDER or MARSH-ROSEMARY

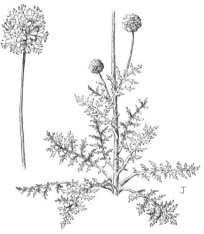

FIGURE 77.   GILIA

FIGURE 78.   ERIASTRUM

SEA-LAVENDER or MARSH-ROSEMARY (*Limonium californicum*), figure 76, is of the Leadwort Family, to which belongs also Thrift (page 46). It is a heavy-stemmed, basally somewhat woody plant, with leaf-blades to about eight inches long and flowering stems to almost two feet high. The flowers are small with whitish sepals and pale violet petals. A European relative of the plant is cultivated for its flower cluster, which becomes papery on aging and can be used for dry bouquets. It is cultivated along the coast and sometimes escapes. The native species is found on beaches and in salt marshes along most of the length of California.

The Phlox Family is for the most part not maritime, but on the coastal strand of central California are a number of subspecies of GILIA (*G. capitata*), one of which, ssp. *Chamissonis*, is shown in figure 77. This Gilia is annual, glabrous to glandular or floccose, one to three feet tall, branched at the base or above, and with dissected leaves. The flowers are in close clusters or heads, the calyx being of five narrow sepals grown together and the corolla of five bluish petals forming a funnel-shaped whole. These lobes of the corolla are narrow to oval. Flowers appear from spring to early summer. See page 102.

Another member of the Phlox Family is ERIASTRUM, a close relative of Gilia, distinguished by unequal calyx lobes and flower heads subtended by bracts. One of the coastal species is *E. densifolium*, figure 78, an erect, much-branched perennial to about one foot high. The leaves are linear and entire or divided into linear lobes. The salverform blue corollas are about one inch long. The typical form grows in sandy places along the coast from Monterey County, California, to Santa Barbara County and a closely related subspecies is in Orange County.

COMFREY (*Symphytum asperum*), figure 79, is a member of the Borage Family (see pages 48, 103), in which the flowers are usually arranged in coiled cymes and the ovary produces four one-seeded nutlets. Comfrey is a native of Asia and is naturalized about Humboldt Bay, California. It has a deep root, stems with recurved hooklike hairs, and bluish flowers about one-half inch long.

FIGURE 79.   COMFREY

Another family with coiled cymes, but with undivided ovaries so that the seeds are borne in small pods, is the Waterleaf Family. Here belongs PHACELIA (*P. Bolanderi*), figure 80. It is perennial from a root-crown, with few stems to two or even three feet long, mostly hairy, and with broad leaves two to four inches long. The corolla is lilac to pale blue, open and spreading. This Phacelia is along the immediate coast from Sonoma County, California, to Oregon, but does get inland sometimes. Flowers appear from May to July. See pages 47, 48, 103.

FIGURE 80.   PHACELIA

The Mint Family, like the two preceding, has the petals united, but here they are organized to form an upper lip and a lower lip. The stems are often square in section and the leaves paired. The fruit produces four one-seeded nutlets. Usually the plants are highly aromatic, as, for example, the Mint, Bergamot, and Pennyroyal. In figure 81 is shown BLACK SAGE (*Salvia mellifera*), a many-stemmed shrub three to six feet high, with green leaves having impressed veins and with heads of pale blue to lavender or whitish flowers about half an inch long. From Contra Costa County, California, southward, it may grow on bluffs overlooking the ocean.

FIGURE 81.   BLACK SAGE

FIGURE 82.
COLLINSIA or CHINESE HOUSES

FIGURE 83.   SYNTHYRIS

FIGURE 84.   CORETHROGYNE

Another family with corollas formed of petals united and two-lipped as are the sages and mints is the Figwort Family (see pages 19, 49–51, 67, 68). A genus of annuals is COLLINSIA, sometimes called CHINESE HOUSES, which is notable in having the middle lobe (petal) of the lower lip folded into a little boatlike structure containing the stamens and pistil. *Collinsia franciscana*, figure 82, is one to two feet tall, often somewhat sticky above, with paired leaves and flowers three-fourths of an inch long. The upper lip is whitish, purple-spotted near the base, while the lower is violet-blue. It is found in brushy and wooded places along the central California coast.

In the same family is SYNTHYRIS (*S. reniformis*), figure 83, of rich coniferous forests from Marin County, California, to Washington. A rather hairy perennial, its leaves are at the base of the plant, its stems to about six inches high, and its blue flowers are about one-third inch long. The capsule is two-lobed.

CORETHROGYNE of the Sunflower Family (pages 20–24, 52–58, 68, 104), with many minute florets or flowers in heads subtended by an involucre of bracts, is a close relative of the Aster and would probably be called that by most of us. It is distinguished however, by having a brushlike microscopic appendage on the style. *Corethrogyne californica*, figure 84, is perennial, more or less permanently white-woolly. Along the coast it has two forms: the one with leaves narrow-oblanceolate and occurring from the Golden Gate to Monterey, the other with leaves spatulate to obovate (egg-shaped with the broad end up) and from Marin County, California, to Coos County, Oregon. The ray-flowers are violet-purple to lilac-pink.

FLOWERS WHITISH TO GREENISH

Section Five

Figure 85 shows SURF-GRASS (*Phyllo-spadix Torreyi*), which with Eel-Grass (*Zostera marina*), belongs to the Eel-Grass Family, a group of seed-bearing plants that are not seaweeds. Both grow submerged in shallow water in bays near the shore and are tossed up on the sand in times of storm. Both have two rows of leaves and minute greenish apetalous flowers arranged on one side of a flattened axis. In Eel-Grass the leaves are one-twelfth to one-third inch wide, in Surf-Grass less than one-twelfth inch. The former ranges from San Diego to Alaska and in Eurasia, the latter from Lower California north.

FIGURE 85.   SURF-GRASS

Another plant which is scarcely a wild-flower, but which is a flowering plant and attracts attention by its odd appearance is ARROW-GRASS (*Triglochin maritima*), figure 86. It is a marsh herb, densely tufted, with stiff narrow leaves and terminal spikes one to two feet high. The minute greenish flowers have six perianth segments. This species ranges in coastal salt marshes from San Francisco Bay northward, but other similar species go as far south as Lower California. The fruit is a cluster of three or six one-seeded carpels.

FIGURE 86.   ARROW-GRASS

CORD GRASS (*Spartina foliosa*), figure 87, is a true grass with its hollow stems and swollen nodes. This species extends along the sandy and marshy shore from Del Norte County, California, to Lower California, but related species go far to the north. Coarse perennials to several feet tall and with strong creeping root-stocks, they are important dune binders. The inflorescence may be as much as a foot long and is somewhat cylindrical, made up of numerous spikelets, each with two basal bracts called glumes and

FIGURE 87.   CORD GRASS

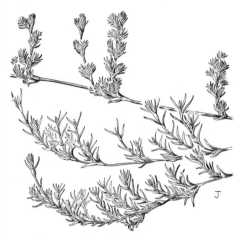

FIGURE 88. MONANTHOCHLOE

FIGURE 89. SALT GRASS

FIGURE 90. POA

a single petalless floret with stamens and pistil.

Among other grasses of coastal salt marshes with striking appearance is MONANTHOCHLOE (*M. littoralis*), figure 88, a spreading wiry-stemmed perennial with clusters of short awl-shaped leaves and short erect branches. The flower-producing spikelets are scarcely evident. This grass occurs from Santa Barbara County, California, to Lower California, in Texas, Florida, Cuba, and Mexico. Staminate and pistillate plants are separate.

Associated with Monanthochloe in salt marshes and often forming large patches is SALT GRASS (*Distichlis spicata*), figure 89. It has a number of technically separated varieties in saline places, ranging from Oregon to southern California. It grows from strong creeping or deeply running rootstocks and has two-ranked leaves four to eight inches long. The spikelets are evident in dense spicate panicles and are more or less green, sometimes purplish. Some forms of this grass are found in salty places inland, even on the desert.

The so-called Blue Grasses are of the genus Poa with two to eight florets in a single spikelet and a basal pair of bracts or glumes. In the common POA of the beach, *Poa Douglasii*, figure 90, we have a low tufted grass spreading by deep-seated rhizomes and with aerial runners to two or three feet long. The stems are stiff, six to sixteen inches long, the leaf-blades stiff and inrolled (note illustration), and the three-to-nine-flowered, pale and tawny spikelets are in dense panicles one to two inches long. This Poa ranges from California's Channel Islands to Puget Sound.

One more unusual and interesting grass of salt marshes and strand is SICKLE GRASS (*Parapholis incurva*), figure 91. In contrast to those grasses mentioned above, it is an annual with slender cylindrical curved spikes in which the one–or–two-flowered spikelets are embedded. It occurs from Oregon to southern California and on the Atlantic Coast, but is apparently native to Europe.

FIGURE 91.   SICKLE GRASS

A family close to the Grasses, but differing from them in tending to have the stems three-sided, usually not hollow, and without swollen hard nodes, is the Sedge Family. Here too the florets are small and arranged in spikelets, but lack the two basal glumes. Cotton-Grass, Tule, Bulrush, Papyrus, and Sedge are in this family. Along our western shore are several, of which two examples are presented here. One is BULRUSH or TULE (*Scirpus robustus*), figure 92, a perennial with horizontal tuber-forming rhizomes and erect sharply triangular stems one and one-half to four and one-half feet tall. At the summit is a tuft of unequal leaves subtending a cluster of ovoid spikelets to an inch long with brownish scales. Each floret has several bristles that represent the modified perianth. It grows in salt marshes of California and Oregon, as well as on the Atlantic Coast and to South America.

FIGURE 92.   BULRUSH or TULE

A second SEDGE is *Carex pansa*, figure 93, a tufted plant from long-creeping rootstocks. It attains a height of about one foot and has few to several crowded spikelets with the stamens at the summit. It is found on beaches and coastal dunes from northern California to Washington.

FIGURE 93.   SEDGE

FIGURE 94.   RUSH

FIGURE 95.   FAIRY BELLS

FIGURE 96.   FRITILLARY

Another family resembling grasses and sedges in its small flowers is the Rush Family, but in it each flower has six small modified greenish perianth parts as well as stamens and pistils and is obviously like a minute lily in structure. In figure 94 is a true Rush (*Juncus acutus* var. *sphaerocarpus*), a perennial forming large tufts two to four feet high and with stout stiff stems. The flowers are two to four in small clusters and up to one-sixth inch long. This plant inhabits coastal salt marshes from San Luis Obispo County, California, to Lower California.

With much larger and more lilylike flowers is FAIRY BELLS (*Disporum Smithii*), figure 95, of the Lily Family (see pages 27, 61, 71, 83). A perennial herb from slender rootstocks, Fairy Bells sends up branched stems to over two feet long with several broad leaves and one to five, mostly whitish flowers in a cluster. The perianth segments are to one inch long. The fruit is a light orange to red berry. The species occurs in moist shaded woods and goes down to the shore at intervals from Santa Cruz County, California, to British Columbia. It flowers from March to May.

In the Lily Family, with its three inner and three outer, usually petallike perianth segments and six stamens and superior ovary is the FRITILLARY (*Fritillaria liliacea*), figure 96. A bulbous plant, with a stem four to fourteen inches high, leaves just above ground level, and one to five bell-shaped whitish flowers with green striations, its flowers are one-half inch or more long. It is found in heavy soil near the coast and ranges from Sonoma County, California, to Monterey County. The flowers appear from February to April.

Another plant in the Lily Family is FALSE LILY-OF-THE-VALLEY (*Maianthemum dilatatum*) shown in figure 97. A low perennial herb with a creeping rootstock, it forms large patches in moist shaded places and grows down on to the actual beach in such spots. The stems are six to fifteen inches high; the small flowers are white and are followed by red berries. The range is from Marin County, California, to Alaska and Idaho. Flowering is from May to June.

FIGURE 97.   FALSE LILY-OF-THE-VALLEY

In the same family is TRILLIUM (*T. chloropetalum*), figure 98. With stout stems a foot or more in height, it bears three large, often mottled leaves and a single sessile flower. This mostly has three greenish petals from one and one-half inches long to almost four inches. There are several forms in the species which may appear near the shore from Santa Barbara County, California, north to Washington and which blooms between February and May.

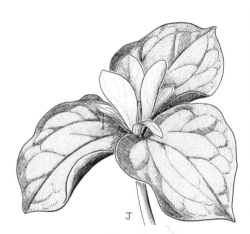

FIGURE 98.   TRILLIUM

For the Orchid Family see page 61. An example to be found on the immediate coast is REIN ORCHID (*Habenaria elegans* var. *maritima*), figure 99. It has fleshy leaves and a dense spike of greenish-white flowers, each with a spur one-half inch or longer. It can be expected on sea bluffs and in similar places from Monterey County, California, to Oregon and blooms from July to September.

The plants discussed under Lily and Orchid Families are Monocotyledons, one of the two main divisions of flowering plants, in which the flowers are built on the plan of three and leaves mostly have parallel veins. In the other great

FIGURE 99.   REIN ORCHID

FIGURE 100. YERBA MANSA

FIGURE 101. WAX-MYRTLE

FIGURE 102. NETTLE

group, the Dicotyledons, the leaves are broad and net-veined, while the flowers are mostly on the plan of four or five. One of these is YERBA MANSA (*Anemopsis californica*), figure 100, of the Lizard-Tail Family. It is highly specialized in having a cluster of many flowers resemble a single one, since the white petallike parts around the base are modified bracts and the individual flowers are minute without sepals or petals, but with six or eight stamens and a pistil. Anemopsis grows in wet places from Santa Clara County, California, south, then east to Texas.

A tall shrub is WAX-MYRTLE (*Myrica californica*), figure 101, of the same genus that is in the northeastern states and produces the bayberry wax so famous for use in candles. As a matter of fact, our western species has small fruits covered with a whitish wax. Wax-Myrtle grows at low elevations along the coast from the Santa Monica Mountains near Los Angeles north to Washington. It is a large shrub to twelve feet or more tall, evergreen, with shining leaves to almost four inches long, and minute flowers in catkinlike clusters. It inhabits mostly canyons and moist slopes.

Quite a contrast to Wax-Myrtle, but also with small greenish flowers in catkins, is the NETTLE (*Urtica holosericea*), figure 102, a perennial herb from underground rootstocks and with stems three to seven feet high. It is covered with bristly hairs that are like small glass bottles that break in the human skin and inject a small quantity of a stinging fluid. It occurs widely in low damp places and can be met along the shore edge as far north as Washington. The small flowers are green and have a deeply parted calyx but no petals.

The Buckwheat Family (see pages 28, 29, 86–87) is remarkable among the broad-leaved plants in having its flowers on the plan of three. One of the conspicuous western genera is CHORIZANTHE (*C. pungens*), illustrated in figure 103. A more or less prostrate annual, it has basal leaves, grayish-hairy stems to one foot long, and dense headlike clusters of minute flowers with a six-parted greenish perianth, each segment of which ends in a recurved spine. It is found in sandy places along the coast from Monterey to San Francisco, California.

FIGURE 103.   CHORIZANTHE

Another group of plants in the Buckwheat Family is Eriogonum or Wild Buckwheat, differing from Chorizanthe in not having spine-tipped perianth-segments. (See pages 28, 29). Among the WILD BUCKWHEATS are many quite attractive species even though the flowers are very small. Such a one is *Eriogonum elongatum* of figure 104, a perennial herb, whitish-woolly throughout, leafy below, and with long leafless branches to two or four feet tall. The small white or pinkish flowers are in cylindrical involucres. It can be found in rocky places along the coast from Monterey County, California, to northern Lower California, as well as farther inland, blooming from August to November.

FIGURE 104.   WILD BUCKWHEAT

Another more strictly coastal WILD BUCKWHEAT is *Eriogonum parvifolium* of figure 105. It is woody, with almost prostrate branches, thinly woolly, especially on the under side of the leaves. Common on bluffs and dunes along the coast, it ranges from Monterey County, California, to San Diego County and bears white flowers tinged with pink. Some flowers can be found through most months of the year.

FIGURE 105.   WILD BUCKWHEAT

FIGURE 106. WILD BUCKWHEAT

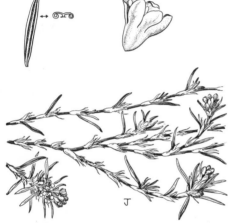

FIGURE 107. KNOTWEED

FIGURE 108. POLYGONUM

A third coastal BUCKWHEAT is *Eriogonum cinereum* shown in figure 106. It is also woody, freely branched, to three or four feet tall, somewhat woolly, with larger leaves and clusters of whitish to pinkish flowers between June and December. It too is found on beaches and on bluffs along the coast, ranging from Santa Barbara County, California, to Los Angeles County. The individual flowers are about one-eighth of an inch long.

KNOTWEED (*Polygonum Paronychia*) of the Buckwheat Family is shown in figure 107. It inhabits the coastal strand from Monterey County, California, to British Columbia and is rather a remarkable species, being quite different from the common Knotweed introduced as a backyard weed from Europe. The present species is a more or less prostrate native perennial from large woody rootstocks, much branched, with papery sheaths at the nodes and with inrolled leaves as shown in the upper lefthand corner of the illustration. The flowers are small, white to pink with green midveins.

Still another POLYGONUM (*P. polystachyum*), figure 108, is included because, although it is not widespread, it is very conspicuous where it does occur, growing to a height of three or four feet, with leaves four to eight inches long, and with large terminal panicles of small white flowers. Found on vacant lots and in coastal marshes in the region of Fort Bragg and Eureka in northern California, it is an introduction from Asia and may well spread more widely in our cool northern coastal region. The flowers appear from June to September.

A perennial with a thick horizontal rootstock and with several erect, slender, simple, smooth stems one to two feet high is KNOTWEED (*Polygonum bistortoides*) shown in figure 109. Commonly thought of as an inhabitant of moist places in the high mountains, it grows also in coastal marshes from Marin County, California, to Alaska and the Atlantic Coast. There in the northern cool climate it sends up its compact spikes of small flowers, each of which flowers, like in the buckwheats in general, has a six-parted perianth, white or pinkish, appearing from June to August.

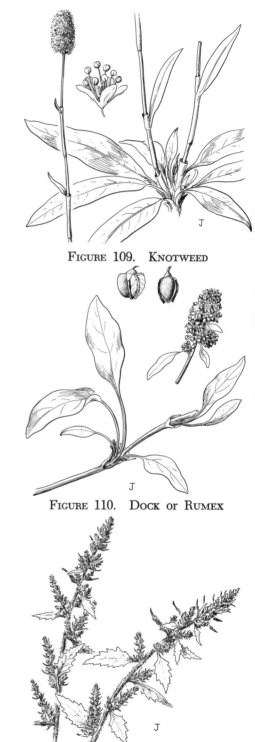

FIGURE 109.   KNOTWEED

In the same Buckwheat Family is the DOCK or RUMEX (*R. crassus*), figure 110, of coastal dunes and rocky ocean bluffs in Los Angeles County and from Monterey County to the state of Washington. It blooms from May to September, the outer sepals of the whitish or greenish flowers being about one-twelfth of an inch long. The three inner sepals or perianth-segments form "valves" that cover the fruit and become one-sixth of an inch, one of them being almost covered by a large callosity. The stems are procumbent to ascending, one to one and one-half feet long, and bear leaves about three times as long as wide.

FIGURE 110.   DOCK or RUMEX

Common in low places, both in the interior and near coastal marshes, is MEXICAN-TEA (*Chenopodium ambrosioides*), figure 111, of the Goosefoot Family (pages 62, 88–90). Like most of that family it has small greenish flowers without petals. It is an ill-smelling plant, annual to perennial, with sprawling stems to a yard long, and with subentire to repand-toothed leaves one to four inches long. The terminal clusters of glandular flowers are quite conspicuous. It is widely spread on the Pacific Coast, naturalized from tropical America.

FIGURE 111.   MEXICAN-TEA

FIGURE 112. SALTBUSH

FIGURE 113. SALTBUSH

FIGURE 114. PICKLEWEED or SAMPHIRE

Figure 112 is of another member of the Goosefoot Family, one of the species of SALTBUSH (*Atriplex patula* ssp. *hastata*); see page 62. In this genus the plant is usually covered with a scurfy coating of inflated, balloonlike hairs, has staminate and pistillate flowers separate, the latter being below the others or on separate plants and in either case being situated between two bracts (see the triangular drawing just to the right of the "J" in the illustration). The Saltbush under discussion grows in salt marshes of the interior and the coast, where it can be found as far north as British Columbia.

Another SALTBUSH is an erect shrub (*Atriplex lentiformis* ssp. *Breweri*), seen in figure 113. It too is grayish-scurfy, attains a height of three to eight feet, and bears leaves one to two inches long. The pair of bracts subtending the pistillate flower is shown in the lower righthand corner and it can be seen how the species in this group differ in these fruiting bracts. This plant is found near coastal salt marshes and on bluffs along the shore and inland, from San Francisco Bay to southern California.

Likewise inhabiting salt marshes and low alkaline places and likewise in the Goosefoot Family is a plant with jointed leafless stems, PICKLEWEED or SAMPHIRE (*Salicornia subterminalis*), shown in figure 114. Its range is from San Francisco Bay to Mexico. It is perennial, but annual species occur with it. The spike to the left shows how the flowers are sunken in a cylindical fleshy axis, green in color, and an inch or more long. This group of plants is world wide.

Yet another member of the Goosefoot Family is SEA-BLITE or SEEPWEED (*Suaeda californica*), figure 115, also of coastal salt marsh and environs. It is interesting how many of this family grow in saline places along the coast or inland, many of them being found in similar habitats in the interior of Asia. This species is perennial, woody at the base, one to almost three feet high, much branched, and with fleshy leaves to an inch or more long. The small greenish flowers have a five-parted calyx and no petals, as is characteristic of the family. Its distribution is from San Francisco Bay to Lower California. It may be smooth or hairy.

FIGURE 115.   SEA-BLITE or SEEPWEED

A last member of this same Goosefoot Family, and one which because of its inconspicuousness should perhaps not be included here, is APHANISMA (*A. blitoides*) of figure 116. My feeling, however, is that many persons buy these little books to help use the large more technical one, *A California Flora*, and that illustrations such as these may help place species that are difficult to handle by keys. This plant is a somewhat succulent annual four to twenty inches tall, with leaves to one inch long, and with small lenticular greenish fruits. Inhabiting coastal bluffs and beaches of southern California and northern Lower California, it is a spring-bloomer.

FIGURE 116.   APHANISMA

Figure 117 is another inconspicuous but locally common maritime plant, SALTWORT (*Batis maritima*) of the Batis Family. Growing on the strand and in salt marshes of southern California, it occurs also on the Atlantic Coast, in the West Indies, and in South America. Prostrate or ascending, woody at the base, the stems become a yard long and bear fleshy leaves one-half inch or longer. The flowers are crowded into catkinlike spikes and have no calyx or corolla, the staminate producing four stamens, the pistil-

FIGURE 117.   SALTWORT

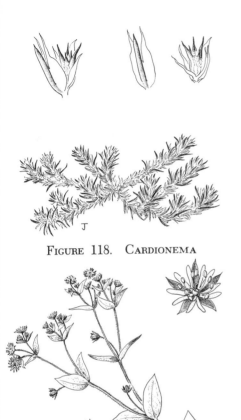

FIGURE 118.   CARDIONEMA

FIGURE 119.   CHICKWEED or STARWORT

FIGURE 120.   PEARLWORT

late one ovary. The pistils coalesce to form a fleshy fruit.

We come now to the Pink Family (page 62) with opposite leaves and often showy flowers. But the species depicted in figure 118, CARDIONEMA (*C. ramosissima*) is far from conspicuous. It is a low tufted perennial, grayish with its short, branched, more or less woolly stems and papery hyaline stipules at the base of the leaves. The middle upper figure shows one of these, two-parted, and with a stiff linear leaf in the middle. The five-parted calyx has unequal sepals ending in short pointed spines. The distribution is remarkable: sandy places along the coast from Washington to Lower California, Mexico, and Chile.

CHICKWEED or STARWORT is the name for the genus Stellaria of the Pink Family and is more representative than is Cardionema. A good example is *Stellaria littoralis* of figure 119, with its deeply two-cleft petals and three styles. A glandular perennial, it has forking stems to twenty inches long, many ovate leaves, and petals about one-fourth inch long. It grows on the strand and adjacent bluffs from San Francisco to extreme northern California, flowering from March to July.

Of the same Pink Family is PEARLWORT (*Sagina procumbens*), a matted perennial or perhaps an annual, shown in figure 120. The prostrate delicate stems root at the node and are one to three inches long. The basal leaves are to three-fourths inch long and bristle-tipped. The flowers have mostly four sepals and petals, sometimes five, and are about one-twelfth inch long. The species is found on moist shaded banks near the beach and on adjacent bluffs, from Point Reyes, Marin County, California, north to British Columbia, also on the Atlantic Coast. It is native of Eurasia.

As a last member of the Pink Family and with opposite, more or less fleshy leaves but with the white petals not cleft as in Starwort, is SAND-SPURREY (*Spergularia macrotheca*), figure 121. At the base of the leaves are papery stipules. The plant is perennial from a heavy branched caudex and a fleshy root, the stems tending to be prostrate and to a foot long. This species of Sand-Spurrey inhabits sea bluffs and is found about salt marshes from British Columbia to Lower California. Somewhat different species grow in alkaline spots in interior valleys.

FIGURE 121.   SAND-SPURREY

NEW-ZEALAND-SPINACH of the Carpet-Weed Family (see pages 13, 30) is an introduced annual that has become naturalized along our beaches and near salt marshes from Oregon south. It is *Tetragonia expansa*, figure 122, with many spreading branches and triangular leaves one to two inches long. The flowers are solitary in the leaf-axils, greenish, with short spreading sepals and no petals. The horned fruit is hard, indehiscent, one-third inch long. The plant came originally from southeastern Asia and from Australasia.

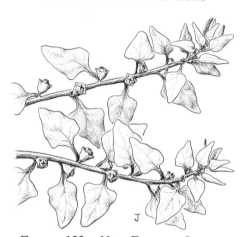

FIGURE 122.   NEW-ZEALAND-SPINACH

Two members of the Portulaca Family (see page 62) are in the genus Montia. One is MONTIA (*M. diffusa*), represented in figure 123, a branched annual two to six inches high, with basal and cauline leaves alike and one to two inches long. The small white flowers have the characteristic pair of fleshy sepals of the family and white or pinkish petals about one-sixth inch long. In woods it comes down to the shore from Marin County, California, to Washington and is sometimes

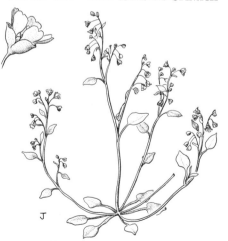

FIGURE 123.   MONTIA

FIGURE 124.   MINER'S-LETTUCE

FIGURE 125.   PEPPERGRASS

FIGURE 126.   WILD RADISH

found almost to the edge of the sand. Flowers are from May to July.

The other Montia shown is commonly called MINER'S-LETTUCE (*Montia perfoliata*), figure 124, remarkable in its pair of connate stem-leaves so different from the basal ones and forming a cup just below the flower cluster. The small white flowers are often recurved in age. The whole plant is fleshy and is edible. Common in much of the West, the species is generally found in shaded places and may occur near the edge of the beach from British Columbia to Lower California. It flowers largely from February to May.

The Mustard Family, with its four-petaled flowers and two-chambered seed-pod, usually has a biting or peppery sap (see pages 15, 32, 63). A white- or green-flowered representative is PEPPERGRASS (*Lepidium oxycarpum*) of figure 125. It is found in saline flats and alkaline valley floors, so that it does occur at the edge of coastal salt marshes of central California. It is a slender-stemmed little annual with leaves one to two inches long and small flowers with or without minute petals. The flat pod is one-eighth of an inch long and has its apex widened into two divergent lobes.

Another member of the Mustard Family is WILD RADISH (*Raphanus sativus*), figure 126. It is a weed of vacant lots and fields, long ago naturalized from Europe, but it grows abundantly on back beaches and adjacent areas. It may cover large areas near the shore and bears its rather showy white or yellowish or purplish flowers with rose or purplish veins. A freely branched annual, it is erect, one to three feet tall, with prominently parted lower leaves. The fruits are characteristically narrowed between the seeds.

BITTER-CRESS or CARDAMINE (*C. angulata*), figure 127, is another member of the Mustard Family and is a forest inhabitant from northern California to British Columbia, coming down to the shore in that cool northern area. It is a perennial from a slender running rootstock, suberect, one to two and one-half feet tall, with angulately lobed leaflets and white petals about one-half inch in length. The spreading elongate seed pods are an inch or so long. Flowers appear largely in May and June.

FIGURE 127.   BITTER-CRESS or CARDAMINE

The Saxifrage Family (see pages 34, 63, 94) is here represented by several plants, one of which BOYKINIA (*B. elata*) is illustrated in figure 128. A slender-stemmed perennial herb, it is erect, one to two feet high, with minute brown, gland-tipped hairs. The lower leaves are one to three inches wide. The flowers are white, one-eighth inch long. It is rather a dainty plant, usually of shaded springy places, occasionally coming out to the shore. Its range is from southern California to Washington.

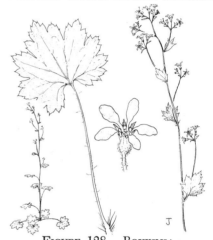

FIGURE 128.   BOYKINIA

Another member of the Saxifrage Family and one that always intrigues me by its finely divided petals is MITRE-WORT (*Mitella ovalis*), figure 129. We have several western species, largely montane, but the one shown here occurs in woods along the coast from central California to British Columbia and may come down to the edge of the shore. It is hairy with retrorse hairs and is a low perennial to about one foot high, blooming largely in April and May. The petals are greenish.

In the same family is ALUM ROOT (*Heuchera pilosissima*), a robust plant

FIGURE 129.   MITREWORT

FIGURE 130.   ALUM ROOT

FIGURE 131.   YERBA DE SELVA

FIGURE 132.   CANYON GOOSEBERRY

shown in figure 130. It is perennial, from an elongate rootstock and has flowering stems one to two feet high and rounded basal leaves one to three inches across. The inflorescence is rather narrow and compact and the petals are pinkish-white and very small. The species occurs on wooded slopes below 1000 feet, from San Luis Obispo County, California, to Humboldt County, flowering from April to June.

In the same Saxifrage Family is a trailing, slightly woody plant, YERBA DE SELVA (*Whipplea modesta*), figure 131. The branches are weak and slender, the leaves opposite and deciduous. The small white flowers are crowded into terminal clusters. Sepals are five to six, thin, erect; petals five to six, white, spreading. The plant is named for Lieutenant Whipple, commander of a government exploring expedition to Los Angeles in 1853 and 1854. It ranges in shaded places in the Coast Ranges from Monterey County, California, to Oregon.

Also in the Saxifrage Family is the genus Ribes, or Currants and Gooseberries. A coastal GOOSEBERRY is *Ribes Menziesii*, figure 132. It has several forms, but one often called CANYON GOOSEBERRY comes out to the coast. It is loosely branched, spiny, shrubby, three to six feet tall, bristly and hairy. The rather firm leaves are one-half to one and one-half inches across and have gland-tipped hairs beneath. The flowers have white petals and purplish sepals and form a globular bristly berry. The range is from southern Oregon to south-central California.

The Rose Family is near the Saxifrage Family in having a sort of tube at the base of the flower, with the sepals and

petals arising from the rim (see pages
34–36, 64, 96). An herbaceous genus is
HORKELIA with white flowers having ten
stamens with dilated filaments (*Horkelia
californica*), figure 133. It is a glandular
perennial, rather pleasantly aromatic.
The main leaves are largely basal, four
to eight inches long, with five to eight or
more pairs of leaflets. The flower tube is
cup-shaped and the white petals about
one-fourth inch long. It is common in
grassy places near the coast of central
and northern California.

Also in the Rose Family is ACAENA (*A.
californica*), figure 134, a rather remarkable plant in having the flower tube
armed with retrorsely barbed prickles.
It is an herb, perennial, one-third to two
feet high, and has deeply cut leaflets silky
beneath. The sepals are green, petals
none. The stamens are dark purple. It is
a plant of the coastal strand and adjacent
bluffs from Sonoma County, California,
to Santa Barbara County.

One of the large groups in the Rose
Family is the BRAMBLE or BLACKBERRY
complex of which *Rubus vitifolius* is illustrated in figure 135. It is a green mound-
builder or trailer or partial climber, with
long stems that tip-root and have many
straightish bristles. The leaves are bright
green above and at most lightly hairy
beneath, in contrast to a closely related
species, *Rubus ursinus*, with duller leaves
which are more or less felted-woolly beneath. Flowers are white in both species
and produce black berries, if not in too
dry a place. They grow along much of
the California coast. *Rubus ursinus* is
an interesting species in that cultivated
berries like Youngberry, Boysenberry,
and Olallieberry have been developed
from it.

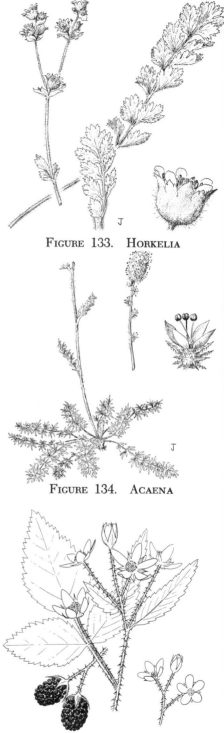

FIGURE 133.   HORKELIA

FIGURE 134.   ACAENA

FIGURE 135.   BRAMBLE or BLACKBERRY

FIGURE 136.   OREGON CRABAPPLE

FIGURE 137.   OSOBERRY

FIGURE 138.   SERVICEBERRY

One group of the Rose Family has applelike fruits with persistent sepal tips at the end. Here belongs OREGON CRABAPPLE (*Malus fusca*), figure 136, a large shrub or small tree with white flowers an inch in diameter and oblong purple-black fruits half an inch long. It is native along the north coast from Sonoma and Napa counties, California, to Alaska. It flowers from April to June.

OSOBERRY (*Osmaronia cerasiformis*), figure 137, has more cherrylike fruits. It is a shrub with simple, entire, deciduous leaves and nodding racemes of fragrant flowers. The fruit is a black stone-fruit with bitter pulp. It is not confined to the coast, but does occur along it in canyons and similar places from Santa Barbara County, California, to British Columbia. This species and the next one are in the Rose Family.

With pome fruits (applelike) is SERVICEBERRY, one species, *Amelanchier florida*, being shown in figure 138. It is a tall slender shrub with erect branches and oblong to rounded leaves over an inch long. The fragrant flowers are in small erect clusters and have white petals. The fruit is purplish-black when ripe and one-half inch in diameter. This species grows along the coast in moist and open places from northern California to Alaska. Its flowers appear from March to May.

A very interesting plant botanically is CROSSOSOMA (*C. californica*), figure 139, in that it combines in its small family characters of the much larger Buttercup Family with several free carpels and of the Rose Family with a more highly developed floral tube. The species shown occurs on Santa Catalina, San Clemente, and Guadalupe islands. It has pure white flowers in early spring. A related species of the deserts is smaller and less conspicuous.

FIGURE 139.   CROSSOSOMA

The Pea Family (pages 16, 36–39, 64, 72) is one of our two largest families with many thousands of species. Among the larger genera is Astragalus, often called Rattleweed, since in those species with inflated pods, the seeds may rattle about, or Locoweed, since some species poison livestock and craze the animals. A RATTLEWEED is *Astragalus Nuttallii* of figure 140. It is a robust perennial, becoming one to almost three feet tall or may be low and matted in windy locations. The flowers are greenish-white, about one-half inch long, while the bladdery pods are one to over two inches long. The species occurs on the mainland coastal strand from Monterey Bay, California, to Point Conception, Santa Barbara County.

FIGURE 140.   RATTLEWEED

Another RATTLEWEED is *Astragalus leucopsis,* figure 141, differing from the preceding species in having its pods on little stalks or stipes above the calyx. The flowers are somewhat larger and the leaflets are not notched at the tips. It grows on sandy bluffs and low hills along the immediate coast, sometimes on shingle banks behind the barrier-beaches. It ranges from Ventura County, California, to northern Lower California, occurring also on several of the Channel Islands.

FIGURE 141.   RATTLEWEED

FIGURE 142.   REDWOOD-SORREL

FIGURE 143.   CROTON

FIGURE 144.   CASCARA

The Oxalis Family is not a family of sea beaches, at least in our part of the world, but the REDWOOD-SORREL (*Oxalis oregana*), figure 142, is a plant of woods close to the shore and may sometimes be found very near it. It has wiry, scaly, branching rootstocks with tufts of leaves and white flowers that are often veined purple. The sepals and petals are five; there are five long stamens and five short. The range is from Monterey County, California, to Washington.

In the Spurge Family, familiar by such plants as Castor-Bean and Poinsettia, the flowers are usually very much reduced, often to single stamens or pistils and grouped in clusters. The pistil has a three-cornered ovary that may become quite conspicuous. Such a plant is CROTON (*C. californicus*), figure 143. It is a more or less hoary perennial one to three feet high, with leaves one-half to one and one-half inches long, no petals, rather many stamens, and the pistils forming capsules one-fourth inch in length. This Croton is found in the interior, but also on the beaches and coastal bluffs from San Francisco to Lower California.

An important northwestern shrub is CASCARA (*Rhamnus Purshiana*), figure 144. It belongs to the Buckthorn Family, together with California-Lilac (pages 40, 73). Again it is not a beach plant, but occurs in nearby forests. However, it can be found along the shore, especially in brushy clearings. It is a deciduous shrub or small tree, having smooth green leaves two to six inches long. The small flowers are followed by rather persistent roundish black berries. It is the inner bark of this species which is important as the source of a well-known laxative. The geographical range is from northern California to British Columbia and Montana.

In the Carrot Family (pages 18, 43, 44, 73, 100) are some small aquatic plants like LILAEOPSIS (*L. occidentalis*), figure 145, a low tufted creeping perennial from long rhizomes. The leaves have no flattened blades, but are reduced to linear structures with transverse partitions. The small umbels of white flowers are an inch or so tall and produce small rounded somewhat corky fruits. Lilaeopsis is found in and near coastal salt marshes from Solano and Marin counties, California, to British Columbia and bears its flowers from June to August.

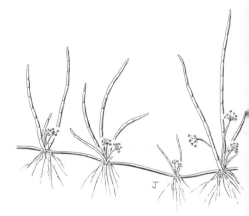

FIGURE 145.   LILAEOPSIS

A larger umbellifer (member of the Carrot Family) of wet places is OENANTHE (*O. sarmentosa*) shown in figure 146. It is a succulent-stemmed perennial to about four feet long, with much divided leaves four to twelve inches long. The small white flowers are in large compound clusters or umbels and have five sepals and five petals. The ovary produces an oblong, often purplish fruit about one-eighth of an inch long. It is found in marshes and sluggish water along and away from the coast, from California to British Columbia and Idaho, and bears flowers from June to October.

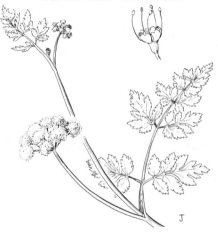

FIGURE 146.   OENANTHE

Likewise in the Carrot Family is ANGELICA (*A. Hendersonii*) shown in figure 147 and found along the coast, both on the strand and neighboring bluffs, from central California to southern Washington. It blooms in June and July, the small white petals being woolly on the back. The plant is stout, perennial, one to almost three feet high, and has large leaves green above and white-woolly beneath. The somewhat winged fruit is shown at the lower left and is about one-third of an inch long.

Another plant with compound umbels,

FIGURE 147.   ANGELICA

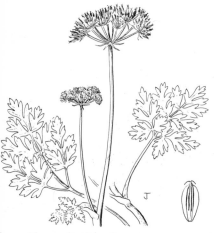

FIGURE 148.   CONIOSELINUM

FIGURE 149.   GLEHNIA

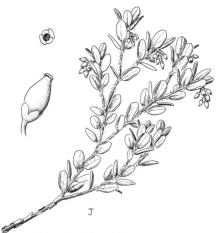

FIGURE 150.   MANZANITA

that is, having small umbels on stalks radiating out from one level, is CONIOSE-LINUM (*C. chinense*), figure 148. It is perennial, stout, mostly branched, glabrous, and from one to five feet tall. The leaves are two to eight inches long, on equally long or longer petioles. This species inhabits ocean bluffs, cold marshes, and the like and ranges from northern California to Alaska, Siberia, and is found also on our Atlantic Coast. The flowers are white, practically without sepals and with no petals. The fruit shown at the lower right is about one-sixth inch long.

The last of the umbellifers shown here is GLEHNIA (*G. leiocarpa*) of figure 149, an almost stemless, more or less fleshy perennial. The sheathing petioles are fairly well buried in the sand and the leaf-blades are one to six inches long, woolly beneath. The plant is otherwise quite hairy. A typical umbelliferous flower with five sepals, five petals, and an inferior ovary and a winged fruit are shown separately. Glehnia grows in beach sand from Mendocino County, California, to Alaska and flowers in May and June.

One of the most important woody genera of the West is MANZANITA of the Heath Family (pages 18, 44, 45, 46). It is characterized by its leathery leaves, its usually reddish bark, its small urn-shaped flowers, and its small usually reddish fruits. A species of acid, often moist places along the coast is *Arctostaphylos Nummularia*, which is sometimes called Fort Bragg Manzanita, but it has also a southern variety, the two forms ranging from the Santa Cruz Mountains, California, to Mendocino County. Figure 150 shows this shrub, which varies from almost prostrate to erect and which has four sepals and four corolla-lobes.

Most species in Arctostaphylos have five-merous flowers, as does another coastal plant, a form of the transcontinental *A. Uva-ursi,* often called BEAR-BERRY or SANDBERRY. It is shown in figure 151. It is a prostrate plant with trailing stems that send up erect branches a few inches high. The white to pinkish corolla is one-sixth inch long and the red berries somewhat larger. It is found in sandy places along the coast from central California to Alaska and east. The more common manzanitas are shrubbier, often quite large, and grow in dry, often rocky places well up into the mountains.

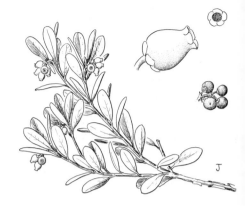

FIGURE 151.   BEARBERRY or SANDBERRY

The so-called SEA-MILKWORT or GLAUX is *Glaux maritima,* a low fleshy perennial depicted in figure 152. It is in the Primrose Family, which includes plants like Primula, Cyclamen, and Shooting Star. This species has a bell-shaped calyx about one-eighth inch long, no petals, and five stamens. The stems are usually less than one foot tall, the leaves about one-half inch long. Glaux is found in coastal salt marsh and on the strand from San Luis Obispo County, California, to Alaska, on the Atlantic Coast, and in the Old World.

FIGURE 152.   SEA-MILKWORT or GLAUX

Another plant of the Primrose Family is STAR FLOWER (*Trientalis latifolia*) of figure 153. It grows in shaded places, chiefly in woods, and may be found adjacent to the very shore from San Luis Obispo County, California, to British Columbia. It is a neat little plant, slender, two to eight inches high, from an underground tuberous rootstock and with a whorl of four to six leaves near the top. The flowers are pinkish-white, about one-half inch broad, and with five to seven sepals and petals. They appear from April to July.

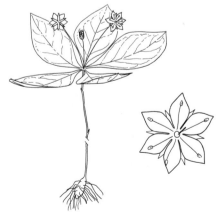

FIGURE 153.   STAR FLOWER

FIGURE 154. ALKALI WEED

FIGURE 155. DODDER

FIGURE 156. LINANTHUS

An inhabitant of saline and alkaline places, hence coming out to the coast, is ALKALI WEED (*Cressa truxillensis*), a low, much-branched, gray perennial shown in figure 154. It belongs to the Morning-Glory Family. It is tufted, woolly-hairy, four to eight inches tall, and has the small white flowers solitary in the upper leaf-axils. It ranges from Oregon to Mexico and Texas and blooms from May to October. The corolla is about one-fourth inch long and has five spreading lobes.

A family close to the Morning-Glory Family is the Dodder Family, in which there is no chlorophyll and the plants are parasitic, twining about their host and sending little knobs or haustoria into it in order to obtain their nourishment. Leafless, or nearly so, they are orange to yellow in color. The DODDER shown in figure 155 is *Cuscuta subinclusa* with rather coarse orange stems and the calyx shorter than the corolla-tube. It is parasitic on California-Lilac, Sumac, and Lemonade Berry, Oak, et cetera, and ranges from Oregon to Lower California. With more slender stems and calyx not shorter than the corolla-tube is *Cuscuta salina*, growing on Cressa, Salicornia, and Chenopodium and ranging as far north as British Columbia.

In the Phlox Family we find no truly beach plants, but some like LINANTHUS (*L. grandiflorus*), figure 156, do get onto the coastal strand and sea bluffs. It is annual, erect, to one and one-half feet high, with leaves cleft to the base into five to eleven linear lobes. The flowers are white to pale lilac and to one inch long. It is primarily a species of central California and blooms from April to July.

The Waterleaf Family, with its flowers in coiling cymes and the ovary forming a capsule, is abundant in western North America. See pages 47, 48, 75. One of its largest constituents is PHACELIA, the blue-flowered species of which are often called Wild-Heliotrope. *Phacelia malvifolia* of figure 157 may be designated as STINGING PHACELIA because of its stiff bristly hairs. It is an annual, erect, one to three feet tall, with broad leaf-blades to about four inches long. The dull white flowers are about one-half inch across. It is found mostly in sandy and gravelly places, hence on back beaches, from Oregon to central California, and blooms from April to July.

An inconspicuous group in the Water-leaf Family is ROMANZOFFIA, with three species on the West Coast. In figure 158 is shown *Romanzoffia Suksdorfii*, a perennial with basal woolly tubers, slender stems to one foot high, and white funnel-form corollas to about one-half inch long. It and the other species are much alike and occur on ocean bluffs and in moist spots in the rocks, from central California to Alaska. Flowering is largely in the spring and summer.

The other common family with coiled cymes is the Borage Family (see pages 48, 75), but its ovary is lobed and forms nutlets. Here belongs WHITE FORGET-ME-NOT (*Cryptantha intermedia*), figure 159. It is an annual, usually stiff-hairy, six to eighteen inches high, with white flowers one-eighth to one-fourth inch broad. The nutlets in each flower are usually four in number and one-seeded. Common in the interior, this species grows along the sea bluffs and on back beaches of southern California and adjacent Lower California. It blooms from March to July.

In the Nightshade Family, well known

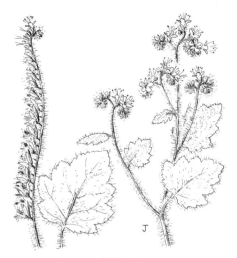

FIGURE 157.   PHACELIA

FIGURE 158.   ROMANZOFFIA

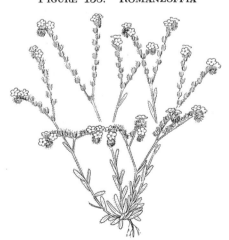

FIGURE 159.   WHITE FORGET-ME-NOT

FIGURE 160. TOBACCO

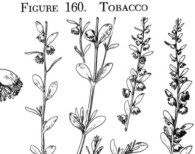

FIGURE 161. POVERTY WEED

FIGURE 162. MALACOTHRIX

for Potato, Eggplant, Tomato, Petunia, and many other common plants is also TOBACCO and among the western species is *Nicotiana Clevelandii*, figure 160. This is a viscid-pubescent annual, one to two feet tall, much branched, with leaves to three inches long. The greenish-white flowers have a rather long corolla-tube and are more than one-half inch across. It is occasional in sandy places on the coastal strand and adjacent bluffs, ranging from Santa Barbara County, California, south.

Our last two plants in this whitish group are in the Sunflower Family (pages 20–24, 52–58, 68, 76) with many minute florets packed into a head that is surrounded by an involucre of bracts. One of these, POVERTY WEED, is *Iva axillaris* of figure 161. It is a low herb, spreading from slender rhizomes, and has leafy stems. The involucre is cup-shaped and surrounds the inconspicuous greenish-white flowers, of which the marginal are fertile and produce seed, while the inner are staminate. It is found in coastal salt marsh, at least in southern California, and east of the mountain ranges as far north as Washington.

MALACOTHRIX (see page 58) is one of the Sunflower Family in which all the florets are strap-shaped and none tubular. A coastal species is the white-headed *Malacothrix saxatilis* shown in figure 162, a plant of sea bluffs. A narrow-leaved form, var. *tenuifolia,* grows on the strand as well as bluffs. A form found mostly on the islands off southern California (var. *implicata*) has the leaves divided into linear segments. The ligules or corollas are about two-thirds inch long and may be more or less rose or purplish. In all forms the blooming season is much of the year.

TREES

Section Six

MONTEREY PINE (*Pinus radiata*), figure 163, is one of the well-known coastal trees of California, although growing in only a few spots in Santa Cruz, Monterey, and San Luis Obispo counties. It can be distinguished by its rather dark green foliage, mostly with three needles in a cluster, the cones asymmetrical and remaining closed and attached to the branches for many years. The wood is light, soft, close-grained, and not strong. A second closed-cone pine is Bishop Pine (*Pinus muricata*), but it is mostly two-needled, the needles four to six inches long. It grows from Humboldt to Santa Barbara counties, California.

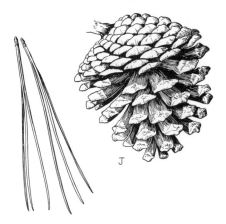

FIGURE 163. MONTEREY PINE

Another coastal pine is BEACH PINE (*Pinus contorta*), figure 164, with mostly two needles to about two inches long and with the cones almost symmetrical, opening and deciduous when mature. It is found on the coastal strand and adjacent bluffs from Mendocino County, California, to Alaska and has a light, hard, strong, brittle, coarse-grained wood occasionally used for fuel.

FIGURE 164. BEACH PINE

TORREY PINE (*Pinus Torreyana*), figure 165, is quite different from the above species in having five needles in a cluster and they are eight to twelve inches long and gray-green. The cones too are large, four to six inches long. Like Monterey Pine it now has a very restricted range as compared to that of prehistoric times, being found only in the region about Del Mar in San Diego County, California, and on Santa Rosa Island. It has a light, soft, coarse-grained wood and, like Monterey Pine, is cultivated in New Zealand, Kenya, and other warmer regions. Like other coastal pines it takes on interesting shapes along the wind-swept coast.

A tree which has been widely in-

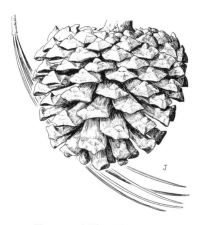

FIGURE 165. TORREY PINE

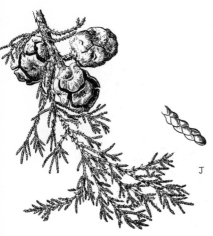

FIGURE 166.   MONTEREY CYPRESS

FIGURE 167.   SITKA SPRUCE

FIGURE 168.   DOUGLAS-FIR

·troduced into cultivation and which is picturesque on the coast is MONTEREY CYPRESS (*Cupressus macrocarpa*), figure 166. The leaves are scalelike, about one-twelfth inch long, bright green. The persistent cones are globose or slightly elongate, an inch or more long, and remain closed for many years. The cone-scales are about eight to twelve in number and bear many seeds. Monterey Cypress is practically confined as a native to the Monterey Peninsula of California. In cultivation, especially in warmer areas away from the coast, it is unfortunately subject to a fungus disease.

Another coastal conifer is SITKA SPRUCE (*Picea sitchensis*), figure 167, a tree of the actual strand and adjacent areas, from Mendocino County, California, north to Alaska. Spruces bear their short needles singly, not clustered, and have the branchlets roughened by the persistent leaf-bases. The bractlets subtending the cone-scales are not exserted or evident. The needles are sessile, usually more or less four-sided, and often sharp-pointed. The cones of this species are oblong, two to four inches long.

DOUGLAS-FIR (*Pseudotsuga Menziesii*), figure 168, which is often confused with spruce, has the branchlets not roughened and the bracts of the cone-scales conspicuously exserted. The needles are flat in cross section and appear two-ranked. It is not primarily a coastal tree, but comes down to the coast and occurs from Monterey Bay, California, north to British Columbia and into the Rocky Mountains. It is the most important lumber tree of North America and its wood is known in the trade as "Oregon Pine." It is perhaps the most commonly used Christmas tree.

WESTERN HEMLOCK (*Tsuga hetero-*

phylla), figure 169, is another forest tree which may reach the coast, where it may be found from northern California to Alaska. Like spruce it has its branchlets roughened by persistent leaf-bases. The branches are slender, more or less pendulous, and bear flat, more or less two-ranked leaves one-fourth to three-fourths inch long. The cones are to about one inch long and have rather thin persistent scales. Hemlock bark has been important in tanning and the rather durable wood of this species is used for construction.

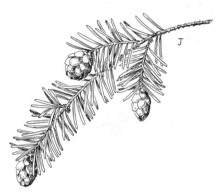

FIGURE 169.   WESTERN HEMLOCK

REDWOOD (*Sequoia sempervirens*), figure 170, is of course one of the famous trees of the world because of its great height and majesty. It has very durable and straight-grained wood. The red, spongy-fibrous bark is conspicuous. The leaves are linear, one-half to one inch long, spreading in two ranks. Redwood Forest is a distinct area in northwestern California, inhabiting the coastal fog belt and it comes out to the shore at the mouth of canyons and gulches. Its range is from the Santa Lucia Mountains of south central California to southwestern Oregon.

FIGURE 170.   REDWOOD

The last conifer for which there is space is GIANT-CEDAR or CANOE-CEDAR (*Thuja plicata*) of figure 171. A forest tree buttressed at the base, it inhabits moist places in the outer Coast Ranges from Mendocino County, California, to Alaska and extends inland to Montana. The leaves are scalelike, mostly about one-eighth inch long, while the cones are about one-half inch long and made up of eight to twelve scales. The wood is light and soft, easily split, so that it is important for shingles and is used also for interior finish.

RED ALDER or OREGON ALDER is *Alnus oregona* and is shown in figure 172. Of

FIGURE 171.
GIANT-CEDAR or CANOE-CEDAR

FIGURE 172.
RED ALDER or OREGON ALDER

FIGURE 173.   GIANT CHINQUAPIN

FIGURE 174.   TANBARK-OAK

the Birch Family, it has deciduous staminate catkins (upper right in the illustration), and more persistent, conelike pistillate catkins. Alder is a deciduous tree and is characterized by few-scaled, long, winter leaf-buds and by toothed leaves. Red Alder differs from our other common tree species, the White Alder (*Alnus rhombifolia*), by having the leaves rusty-pubescent beneath, inrolled on the margins, and by its narrowly winged seeds. The range is in damp places from Santa Cruz County, California, to Alaska.

In the Beech Family, with Oak and Chestnut, is GIANT CHINQUAPIN (*Castanopsis chrysophylla*), figure 173. It is a tall tree with heavily furrowed bark, coriaceous leaves two to six inches long and staminate catkins as shown upper right. The pistillate structures produce burs with long spines and inclose one to three nuts to about one-half inch long. This is a forest tree coming out to the beach and ranging from Mendocino County, California, to Washington. The fruit matures in the second season.

Like the Chinquapin in its long staminate catkins with their strong odor and like the Oak in its acorn is TANBARK-OAK (*Lithocarpus densiflora*), figure 174. It is evergreen with a narrow conical crown. The acorn cup has slender spreading scales and is quite different from that of the Oak. In North America we have a single species throughout the Coast Ranges from Ventura County, California, to southern Oregon, but southeastern Asia has about one hundred species. In the mountains our species has a dwarf shrubby form.

The genus of the Oak has a tremendous number of species, one of the most common in California being the COAST

Live Oak or Encina (*Quercus agrifolia*), figure 175. It is a broad-headed tree and takes on very picturesque shapes in age. The rather harsh leaf blades are characteristically inrolled slightly at the edge; the acorns are long and pointed. Common over much of California between the Sierra Nevada or more southern mountains and the coast, the species reaches the shore itself, especially at the mouth of water courses. In much of central California it is often festooned with a netlike gray lichen sometimes wrongly called "Spanish-Moss."

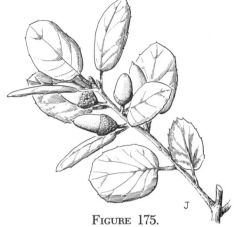

FIGURE 175.
COAST LIVE OAK or ENCINA

FIGURE 176.
CALIFORNIA-BAY or CALIFORNIA-LAUREL

Another widely distributed tree in California and extreme southwestern Oregon, where it is called "Myrtle," is a member of the Laurel Family, CALIFORNIA-BAY or CALIFORNIA-LAUREL (*Umbellularia californica*), figure 176. Like the other trees enumerated here, it actually reaches the shore, although primarily a forest tree. It is pungently aromatic, has stiffish or rubbery leaves and small yellow-green flowers. It is often used for seasoning as a substitute for the true Bay. The wood is hard, strong, and takes a high polish, so that it is very good for turning. It blooms from December to May.

Maples have opposite, broad, usually lobed but sometimes compound leaves. The small flowers are variously clustered and have four to nine stamens and a two-styled pistil which forms two-winged fruits united below and called samaras. The species most apt to appear on the shore is VINE MAPLE (*Acer circinatum*), figure 177, which ranges from northern California to British Columbia. It is vine-like and reclining, with slender twigs and leaves two to five inches across. Canyon Maple (*Acer macrophyllum*) is stouter and has much larger leaves.

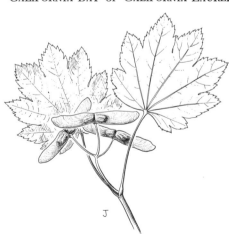

FIGURE 177.   VINE MAPLE

COASTAL COUNTIES OF WASHINGTON, OREGON, AND CALIFORNIA

INDEX TO COLOR PLATES

(References are to plate numbers)

113

INDEX

(References are to page numbers)